AQA (A) GCSE

Revision Guide

Geography

Jane Ferretti

Brian Greasley

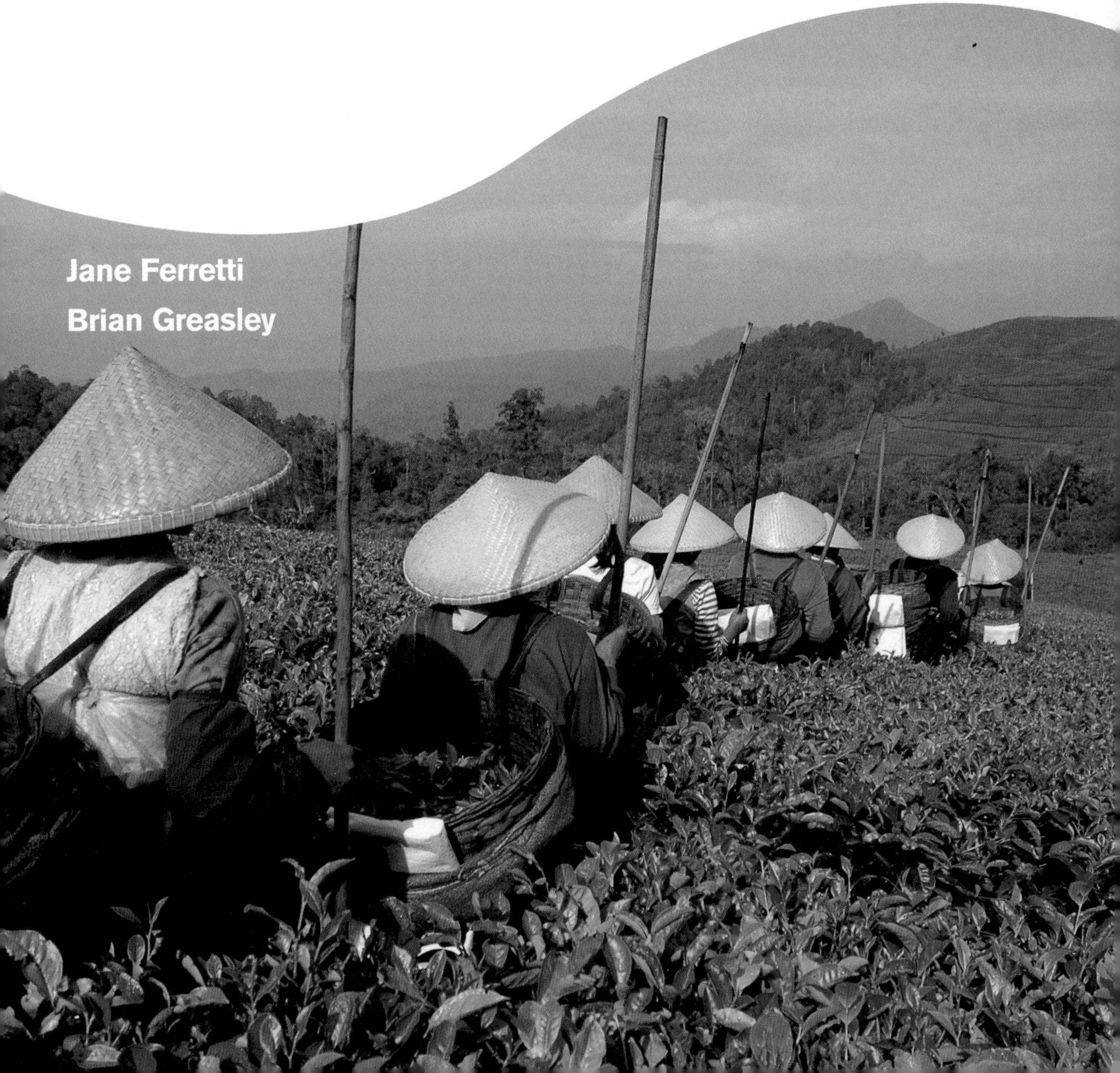

Philip Allan Updates, an imprint of Hodder Education, an Hachette UK company, Market Place, Deddington, Oxfordshire OX15 0SE

Orders

Bookpoint Ltd, 130 Milton Park, Abingdon, Oxfordshire OX14 4SB
tel: 01235 827827
fax: 01235 400401
e-mail: education@bookpoint.co.uk

Lines are open 9.00 a.m.–5.00 p.m., Monday to Saturday, with a 24-hour message answering service. You can also order through the Philip Allan Updates website: www.philipallan.co.uk

ISBN 978-1-4441-1045-6

First printed 2011
Impression number 5 4 3 2 1
Year 2015 2014 2013 2012 2011

Printed in Italy

P01801

Contents

About this book

Revision is vital for success in your GCSE examination. No one can remember what they learnt up to two years ago without a reminder. To be effective, revision must be planned. This book provides a carefully planned course of revision – here is how to use it.

The book	*The route to success*
Contents list	**Step 1** Check which topics you need to revise for your examination. Mark them clearly on the contents list and make sure you revise them.
Revision notes	**Step 2** Each section of the book gives you the facts you need to know for a topic. Read the notes carefully, and list the main points.
Key words	**Step 3** Key words are highlighted in the text. Learn them and their meanings. They must be used correctly in the examination.
Case study	**Step 4** Each section has a case study. Learn one for each topic. You could do this by listing the main headings with key facts beneath them on a set of revision cards. If your teacher has taught different case studies, use the one you find easiest to remember.
Test yourself	**Step 5** A set of brief questions is given at the end of each section. Answer these to test how much you know. If you get one wrong, revise it again. You can try the questions before you start the topic to check what you know.
Examination questions	**Step 6** Examples of questions are given for you to practise. Notice the higher-tier question at the end for those entered for the higher tier. The more questions you practise, the better you will become at answering them.
Exam tips	**Step 7** The exam tips offer advice for achieving success. Read them and act on the advice when you answer the question.
Key word index	**Step 8** On pp. 142–144 there is a list of all the key words and the pages on which they appear. Use this index to check whether you know all the key words. This will help you to decide what you need to look at again.

Command words

All examination questions include **command** or **action** words. These tell you what the examiner wants you to do. Here are some of the most common ones.

- **List** – usually wants you to provide a list of facts.
- **Describe** – requires more than a list. For example, you are expected to write a description of the pattern of population in Wales, but not to give any explanation for it.

- **Explain**, **give reasons for** or **account for** – here the examiner is expecting you to show understanding by giving reasons, and to do more than describe, for example, the pattern of population in Wales.
- **Suggest** – the examiner is looking for sensible explanations, using your geographical knowledge, for something to which you cannot know the actual answer – for example, 'Suggest reasons for the location of the factory in photograph A'.
- **Compare** – the best candidate will not write two separate accounts of the factors to be compared, but will pick several points and compare them one at a time. Useful phrases to use are 'whereas', 'on the other hand', 'compared to'.

Do you know?

- The exam board setting your paper?
- What level or tier you will be sitting?
- How many papers you will be taking?
- The date, time and place of each paper?
- How long each paper will be?
- What the subject of each paper will be?
- What the paper will look like? Do you write your answer on the paper or in a separate booklet?
- How many questions you should answer?
- Whether there is a choice of questions?
- Whether any part of the paper is compulsory?

If you don't know the answer to any of these questions as the exam approaches – ask your teacher!

Revision rules

- Start early.
- Plan your time by making a timetable.
- Be realistic – don't try to do too much each night.
- Find somewhere quiet to work.
- Revise thoroughly – reading is not enough.
- Summarise your notes, make headings for each topic, and list the case study examples.
- Ask someone to test you.
- Try to answer some questions from old papers. Your teacher will help you.

If there is anything you don't understand – ask your teacher.

Be prepared

The night before the exam

- Complete your final revision.
- Check the time and place of your examination.
- Get ready your pens, pencil, coloured pencils, ruler and calculator (if you are allowed to use one).
- Go to bed early and set the alarm clock!

On the examination day

- Don't rush.
- Double check the time and place of your exam and your equipment.
- Arrive early.
- Keep calm – breathe deeply.
- Be positive.

Examination tips

- Keep calm and concentrate.
- Read the paper through before you start to write.
- If you have a choice, decide which questions you are going to answer.
- Make sure you can do all parts of the questions you choose, including the final sections.
- Complete all the questions.
- Don't spend too long on one question at the expense of the others.
- Stick to the point and answer questions fully.
- Use all your time.
- Check your answers.
- Do your best.

Unit 1

Physical geography

Plate tectonics

The Earth's **crust** is made up of pieces like a jigsaw, called **tectonic plates**.

The plates 'drift' or move because the heat in the rock below the crust sets up convection movements (rather like the movement you see in a pan of soup when it is heated).

The Earth

Crust (6–80 km thick)

Mantle, approx 3,000°C (2,800 km thick)

Outer core of nickel and iron, molten liquid (3,500 km thick)

Inner core, solid (1,200 km thick)

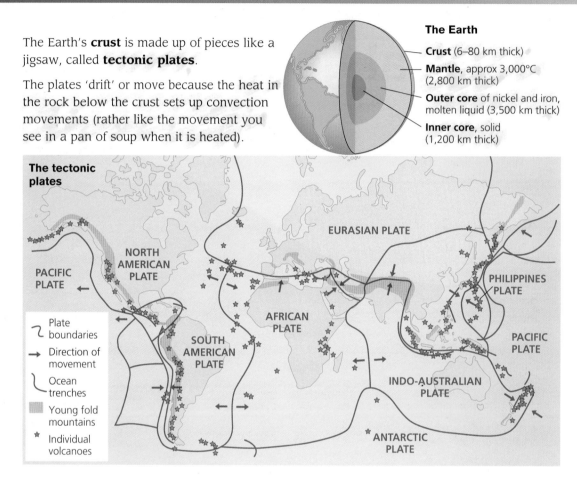

The tectonic plates

EURASIAN PLATE

NORTH AMERICAN PLATE

PACIFIC PLATE

PHILIPPINES PLATE

AFRICAN PLATE

SOUTH AMERICAN PLATE

PACIFIC PLATE

INDO-AUSTRALIAN PLATE

ANTARCTIC PLATE

Plate boundaries

Direction of movement

Ocean trenches

Young fold mountains

Individual volcanoes

The movement of tectonic plates

The plates are moving all the time. At the margins (edges) they either move apart, together or slide past each other.

Plate movement

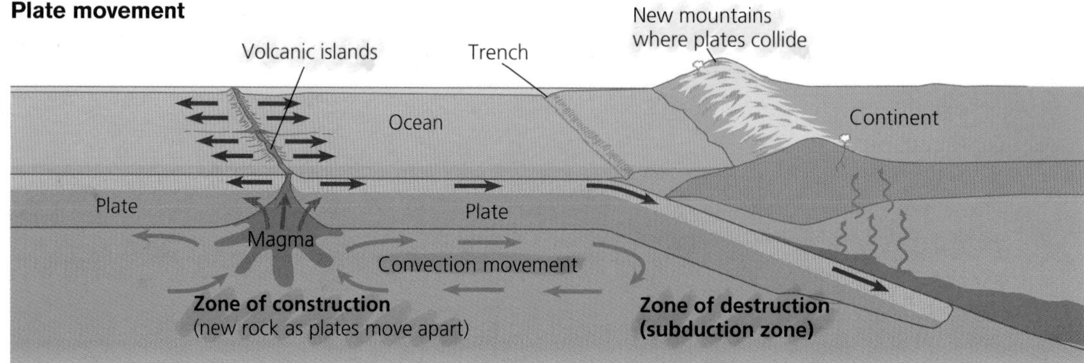

Types of plate margin

Plates meet at plate margins. There are four different types.

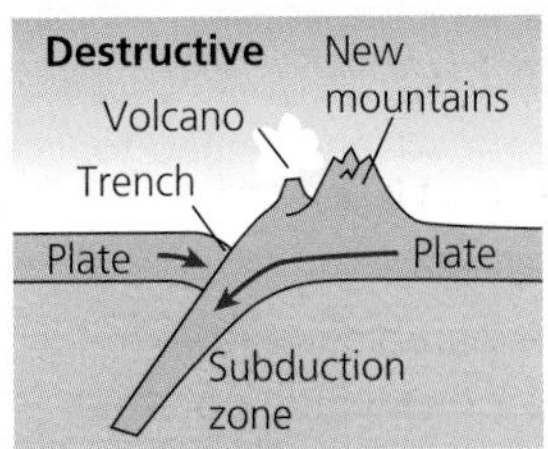

At a **destructive plate margin** one plate slides beneath another as they collide, a process known as **subduction**. The bottom plate crumples creating new mountains.

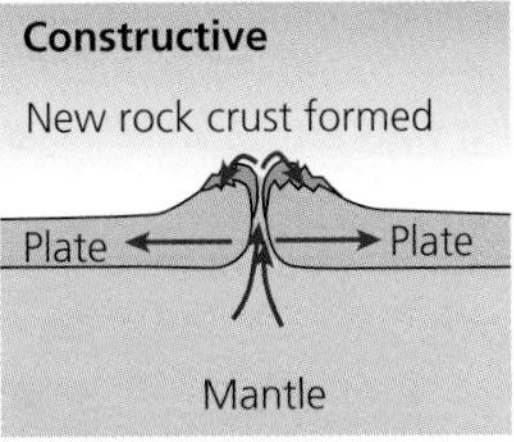

At a **constructive plate margin** the plates are moving apart. Molten rocks from the mantle below spread out and harden forming a ridge of new rock.

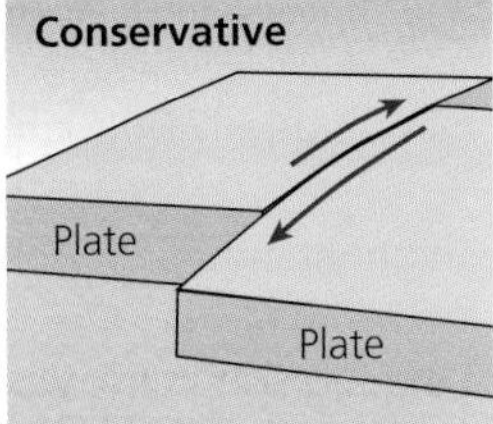

At a **conservative plate margin** the plates slide past each other. Pressure builds up until they 'jerk' causing earthquakes.

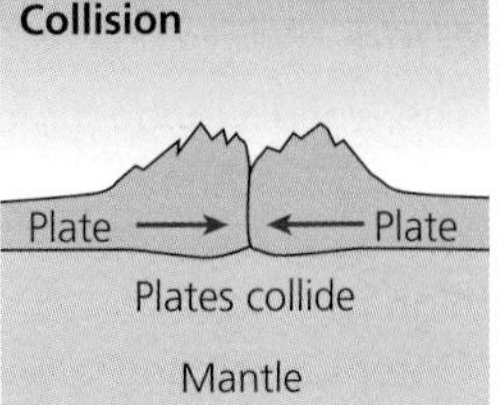

A **collision plate margin** is where two plates collide and are crushed against each other. They are pushed upwards, forming new mountains.

The difference between oceanic and continental plates

The differences between **oceanic** and **continental plates** are:

- the oceanic plates are younger than the continental plates
- the oceanic plates are being renewed and destroyed at the **zones of construction and destruction**. The continental plates cannot be renewed or destroyed
- the continental plates are less dense than the oceanic plates
- the oceanic plates can sink whereas the continental plates cannot

Unique landforms at the plate margins

Three unique landforms occur at the plate margins. They are **ocean trenches**, **fold mountains** and **volcanoes**. (See map opposite.)

Ocean trenches

Deep ocean trenches are formed offshore at destructive plate margins (subduction zones) where the denser oceanic plate dives beneath the less dense continental plate. These are the deepest parts of the ocean such as the Japan Trench, the Peru–Chile trench and the Mariana Trench (over 10,920 m deep).

Fold mountains

When tectonic plates collide, huge forces cause the rocks to be folded and buckled, forming mountains like the Andes in South America (a subduction zone at a destructive plate margin) and the Himalayas in Asia (a collision zone).

An upfold is called an **anticline**. A downfold is called a **syncline** (sink = down)

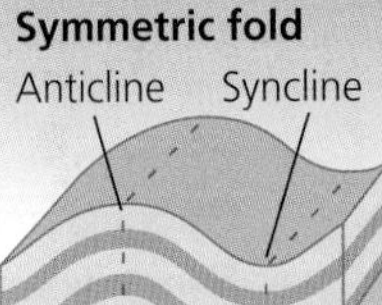

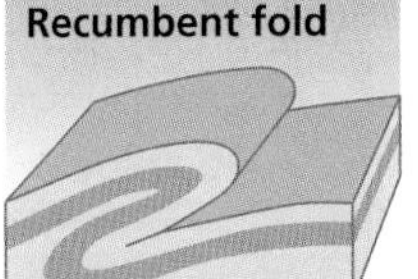

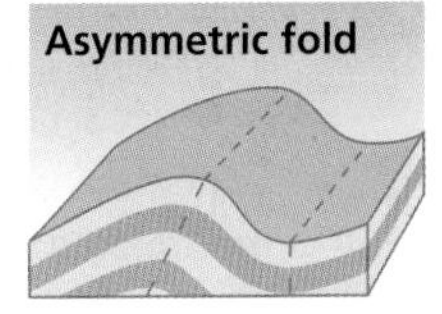

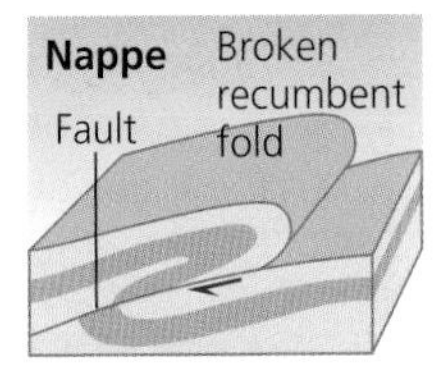

Types of fold

Key words

crust	subduction zone	oceanic plate	ocean trench
tectonic plate	constructive plate margin	continental plate	fold mountain
mantle	conservative plate margin	zone of construction	volcano
outer core	collision plate margin	zone of destruction	anticline
inner core			syncline
destructive plate margin			

Case study: *one range of fold mountains*

The Andes Mountains of South America

Factfile

- The world's longest mountain range – 7,000 km long with an average width of 200 km.
- Average height 4,000 m – highest mountain Aconcagua is 6,692 m high.
- Cotopaxi volcano is one of the highest in the world.
- Where the north–south ranges widen in Peru and Bolivia there are bleak high plateaus called the Altiplano.

Farming

- Farmed land limited to the valley bottoms and terraced slopes in favourable areas.
- Crops include potatoes and maize – subsistence crops grown for the farmer's family – and crops such as soybeans, tobacco, coffee and cotton to sell.
- Animals such as llamas and alpacas are kept on the high plateaus for transport as pack animals and for their meat, milk and wool.

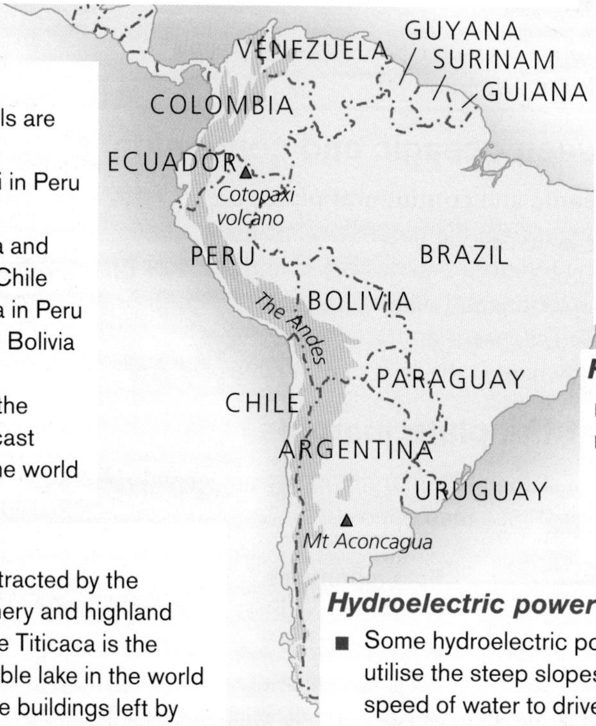

Mining

A range of minerals are mined including:

- silver at Potosi in Peru
- copper at Chuquicamata and Escondida in Chile and Toquepala in Peru
- tin in Peru and Bolivia
- gold in Peru – Yanacocha is the largest open-cast gold mine in the world

Formation

- Range of fold mountains.
- Caused by the subduction of the Nazca tectonic plate beneath the South American Plate.

Tourism

- Tourists are attracted by the mountain scenery and highland lakes, e.g. Lake Titicaca is the highest navigable lake in the world
- The ruins of the buildings left by the Incas at Machu Pichu and Cusco attract thousands of visitors each year

Hydroelectric power

- Some hydroelectric power is being developed to utilise the steep slopes which provide a good head speed of water to drive the turbines, and narrow valleys to dam the flow of water and provide a steady supply.
- The Yuncan Project Dams are in northeast Peru.

How people adapt

- The steep relief and parallel north–south ridges limit communications, particularly east to west across the mountains.
- Many of the passes are very high, e.g. the pass between Arequipa and Puno in Peru is 4,470 m high.
- The major transport route between Chile and Argentina is the Uspallata pass 3,800 m high, a highway follows the pass and the railway tunnels beneath it.
- People have adapted to these harsh highlands by living in the valleys and by keeping animals that are suited to the environment, and by terracing the hillsides to gain additional space.

Volcanoes

Distribution of volcanoes

The majority of volcanoes occur at the tectonic plate margins. Particularly noticeable is the concentration around the Pacific Ocean which follow the plate margins of the Pacific Plate – known as the 'Ring of Fire'.

Why volcanoes occur

Molten rock known as **magma** can escape when pressure builds up below the Earth's surface, particularly at plate margins. When the magma reaches the surface it is known as **lava**.

Volcanoes may erupt very explosively at destructive plate margins releasing enormous amounts of lava, ash and steam. Such volcanoes, made up of ash and lava, are known as **composite volcanoes**.

Sometimes the lava is more liquid and flows more gently to the surface through cracks in the crust at constructive plate margins. These volcanoes are known as **shield volcanoes**. A shield volcano may also form when a plate is over a particularly hot part of the mantle known as a **hot spot**. Such volcanoes may occur away from the plate margin.

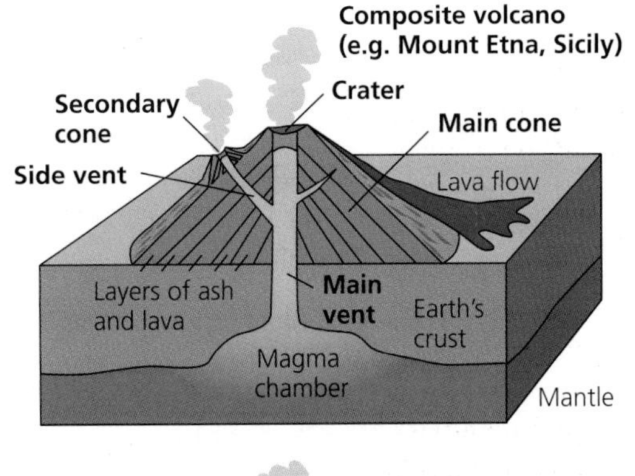

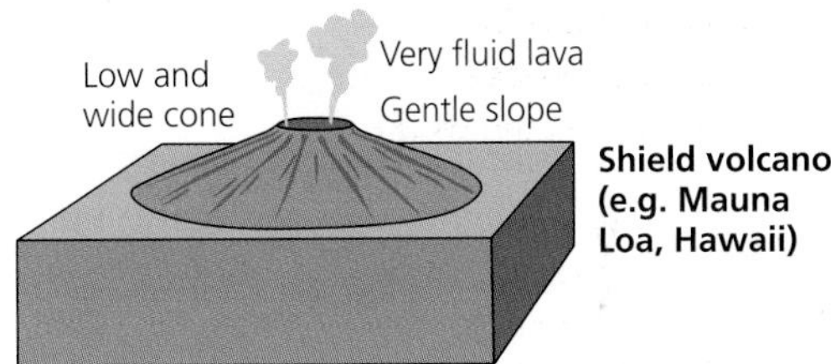

Volcano facts

- When a volcano erupts violently it throws ash and **volcanic bombs**, which are pieces of rock, into the air.
- Hot gases, ash and steam can form **pyroclastic flows** which move very fast and can cause tremendous damage.
- When a volcano has been dormant for some time the solidified magma in the vent acts as a plug. When the volcano erupts the **plug** is blown out, often blowing off the top of the cone and leaving a very large crater known as a **caldera**.
- Mudflows called **lahars** are formed when hot ash melts snow and ice or falls into rivers. They move very fast and are very destructive.

Key words

magma
lava
composite volcano
shield volcano
hot spot
volcanic bomb
pyroclastic flow
plug
caldera
lahar
main cone
secondary cone
crater
main vent
side vent
active volcano
dormant volcano
extinct volcano

Types of volcano

- **Active volcanoes** have erupted recently and are likely to erupt again.
- **Dormant volcanoes** have not erupted for a long time but may erupt again.
- **Extinct volcanoes** are unlikely to ever erupt again.

Reducing the impact of a volcanic eruption

Monitoring and prediction

Effective monitoring and prediction enable authorities to warn people living nearby to evacuate the area. A number of indicators warn of impending eruption:

- the frequency of earth tremors
- gravity measurements of the movement of magma inside the volcano
- the build up of magma causes ground temperatures to rise
- emissions of gases such as sulphur dioxide rise

Reducing the impact

- **Evacuation** Warning and evacuation procedures to reduce the loss of human life.
- **Mapping** Paths of old lava flows and mudflows can show areas of risk.
- **Engineering** Minor lava flows can be diverted by bulldozing walls to turn the flow away from villages. In Iceland seawater is sprayed onto lava flows to solidify them.

You need to know a case study of a volcanic eruption, e.g. Mount Etna, 2002.

A supervolcano: Yellowstone, USA

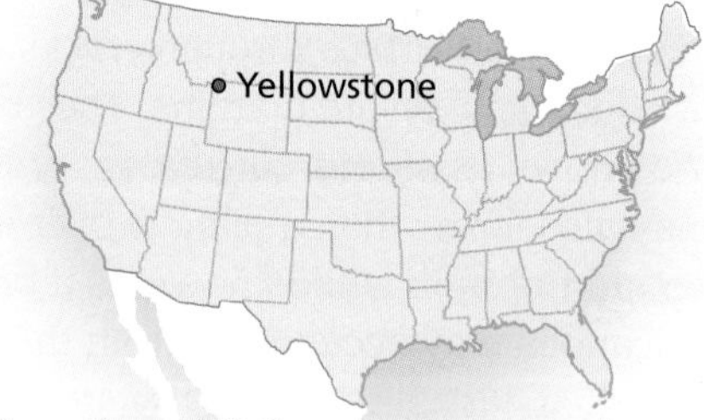

Location of Yellowstone supervolcano

A **supervolcano** is a gigantic volcano. It will not have a cone like a normal volcano but will sit in a large depression called a caldera**.** The caldera at Yellowstone is 55 km by 72 km.

Yellowstone lies on top of a hot spot where light molten magma rises towards, and collects in, a store beneath the surface. Volcanic eruptions sometimes empty their store so quickly that the land collapses into the emptied chamber and leaves a caldera.

The geysers for which Yellowstone is famous are caused by water becoming superheated on the rocks beneath the surface and steam blasting into the air through vents or cracks in the surface.

The likely effects of a Yellowstone eruption

An explosive super-eruption of Yellowstone supervolcano would be devastating and would cover large areas of North America in ash which would affect farming as crops would fail and animals would die from the ash. Communications would be disrupted, buildings over a wide area would be damaged and it would cause great loss of life. Global climates would change as a result of the ash in the atmosphere.

Earthquakes

Earthquakes and volcanoes are examples of natural hazards, they tend to occur on the plate margins. The major earthquake zones are around the Pacific Ocean and in mountainous regions such as the Himalayas, Central China and Iran. Not all earthquakes occur on land – many happen beneath the sea where they can cause enormous tidal waves known as **tsunami**, which can devastate coastal areas.

Earthquakes are usually the result of plate movement, which is *not* smooth. The strain builds up along a fault line between two plates until they move, causing earthquakes.

When shockwaves reach the surface they cause the ground to shake from side to side, causing damage

Where plates collide, rock layers are forced upwards to produce mountain ranges

Epicentre: point on the Earth's surface directly above the hypocentre and where shock waves are strongest and most damage occurs

Shockwaves

When plates slip under pressure they release shockwaves

Bedrock

Focus or **hypocentre**: point of the earthquake's origin

Focus and epicentre

The point where the earthquake starts below the earth's surface is known as the **focus** or **hypocentre**. The point directly above the focus on the Earth's surface is known as the **epicentre.**

Measuring earthquakes

The magnitude of an earthquake is recorded by an instrument called a **seismometer**. It measures the height of the shock waves on the **Richter scale**. Each point on the scale is 10 times greater than the one below. This means that an earthquake with a score of 7 is 10 times more powerful than one with a score of 6. The seismometer records the earthquake vibrations with a pen on a sensitive arm, marking zigzag lines on a drum of paper.

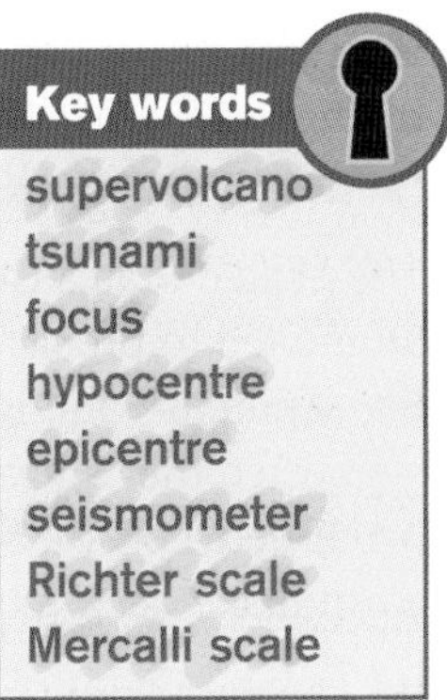

The **Mercalli scale**, which has been largely replaced by the Richter scale, measures the strength of an earthquake by its observed effects on buildings. So 1 on the Mercalli scale would be detectable only by seismographs and there would be no damage at all, while 12 would be catastrophic, with no buildings left standing.

Precautions against earthquakes

Individuals

- Prepare an emergency pack including water, food, blankets, first-aid kit, radio and torch.
- During and after the earthquake, shelter under a table or bed, avoid stairways.
- Turn off gas, water and electricity.
- After the earthquake, move to open ground.

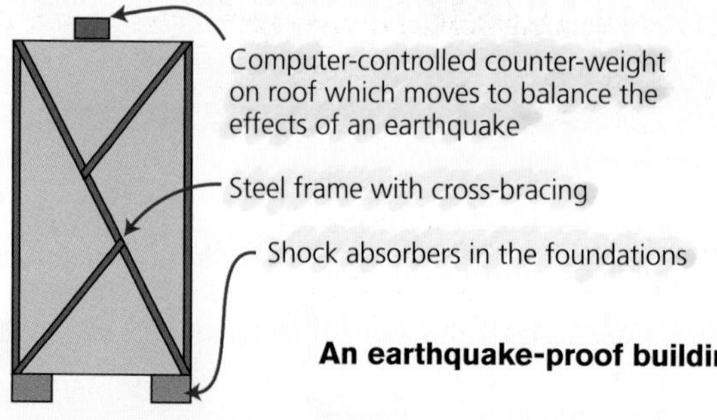

An earthquake-proof building

Authorities

- Monitor the hazard so people can be warned.
- Have emergency supplies ready.
- Make plans for shelter, food and water supplies, and for emergency services such as fire brigade, police, ambulance and hospital services.
- Plan to broadcast information for people affected.

Long-term planning

- Long-term planning is more likely to be sustainable.
- Ensure road and rail communications are built to reduce the effect of earthquakes.
- Ensure all new buildings are earthquake proof.
- Provide education and advertising so all people know what to do in the event of an earthquake.

Predicting earthquakes

A great deal is known about where earthquakes are likely to occur but it is impossible to predict exactly where or the day or month when an earthquake will happen in a specific location.

The only really successful prediction was in China in 1975. The prediction was based on changes in the height of the land, levels of underground water, animal behaviour and, most tellingly, a sudden increase in the number of 'foreshocks'.

The United States Geological Survey attempted to predict activity on the San Andreas Fault in California. They suggested that an earthquake would occur between 1988 and 1992 – an earthquake occurred in 2004.

You need to know case studies of an earthquake in a rich country (e.g. Kobe, 1995) and one in a poor country (e.g. Haiti, opposite).

Impact of earthquakes

The effect or impact of an earthquake depends on the magnitude of the quake and the density of the population and human activity, such as industry, in the area.

People and authorities in richer areas are generally more prepared than those in poorer areas. Their greater wealth enables them to develop earthquake-proof buildings, more effective emergency services and a speedier response.

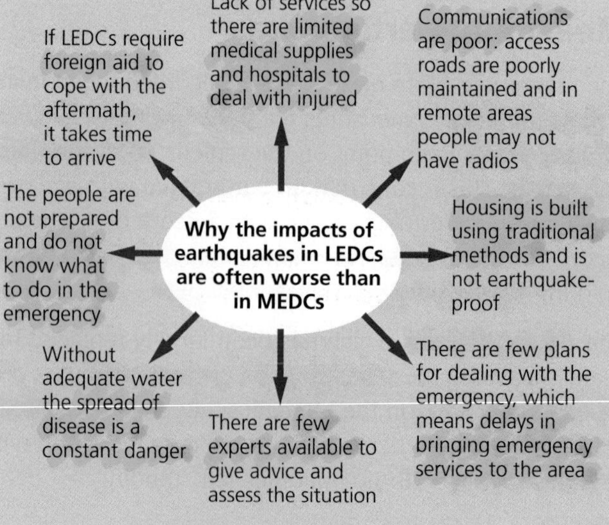

Why the impacts of earthquakes in LEDCs are often worse than in MEDCs
Lack of services so there are limited medical supplies and hospitals to deal with injured
Communications are poor: access roads are poorly maintained and in remote areas people may not have radios
Housing is built using traditional methods and is not earthquake-proof
There are few plans for dealing with the emergency, which means delays in bringing emergency services to the area
There are few experts available to give advice and assess the situation
Without adequate water the spread of disease is a constant danger
The people are not prepared and do not know what to do in the emergency
If LEDCs require foreign aid to cope with the aftermath, it takes time to arrive

Case study: *an earthquake in a poor part of the world*

Haiti, 2010

Factfile

- Earthquake hit on 12 January 2010.
- Measured 7.0 on the Richter scale.
- Epicentre near the town of Léogâne, 25 km (16 miles) west of the capital Port-au-Prince.
- Haiti is an extremely poor country ranked 149 out of 182 on the Human Development Index (see page 113).
- Haiti lies at boundary of two tectonic plates, the Caribbean Plate and the North American Plate.

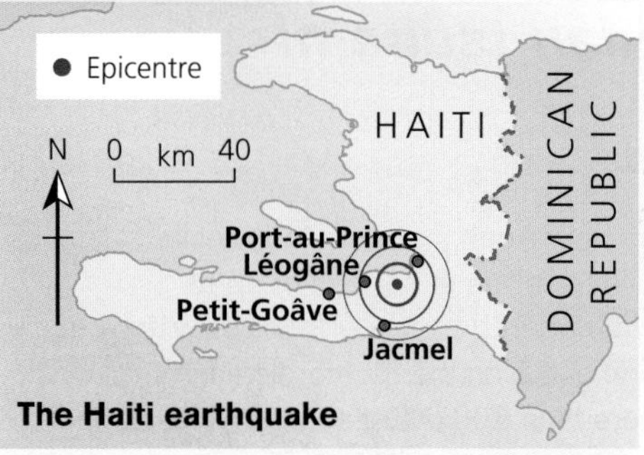

The Haiti earthquake

Primary effects

- Violent shaking of the ground and buildings and in the 12 days afterwards 52 aftershocks measuring 4.5 or greater on the Richter scale were recorded.
- Estimated 230,000 people killed, 300,000 injured and over 1 million made homeless out of a population of 10 million.
- 250,000 dwellings destroyed, electricity supplies disrupted, the international airport and main port damaged.
- Public telephone system stopped operating and no signal from mobile phones and no internet.

Secondary effects

- Main roads blocked for 10 days after the shock.
- Hospitals and medical facilities were damaged or destroyed delaying medical care for injured.
- Damage to the airport and port delayed the arrival of medical supplies.
- Construction standards in Haiti are low – many buildings collapsed including the Presidential Palace, the Cathedral, the City Hall and the offices of the World Bank.
- Two million Haitians live as squatters and many were made homeless.

Initial response

- Initially there was confusion. Despite previous disasters there was little forward planning or preparation and, as a result, aid was slow to arrive.
- People slept in the streets – there were not enough tents.
- Because provisions of water, food and medical supplies were slow to arrive there was widespread looting and some violence.
- As there had been few drills, people were unprepared.
- Services were overwhelmed and corpses buried in rubble began to decompose – mass graves were dug.
- Aid (tents, field hospitals and food) began to arrive during the days following the quake, although chaotically at first. Many countries and people around the world pledged money.

Long-term response

- One year after the earthquake many were still living in tents. Aid agencies warned of a further humanitarian disaster if shelter and sanitation were not rapidly put in place.
- The Haitian economy is in ruins – 1 in 5 jobs have been lost.
- Before the earthquake, Port-au-Prince could not provide for all the people who migrated from the countryside to find work. Since the earthquake it has become impossible – many have returned to their villages.
- The UN and international community recognise that Haiti will need massive support to recover and that recovery will be slow.

Case study: *a tsunami*

The Asian tsunami

Factfile

- Tsunami struck on 26 December 2004, its epicentre near Aceh, northern Indonesia.
- Tidal wave travelling at speeds of up to 800 km/h triggered by an earthquake with a magnitude of 9.1 on the Richter scale.
- Killed more than 200,000 people in coastal areas in 13 countries; 128,000 died in Indonesia alone.
- £7 billion pledged by people around the world.
- Destruction high as many of the affected areas had dense populations living in sub-standard housing.

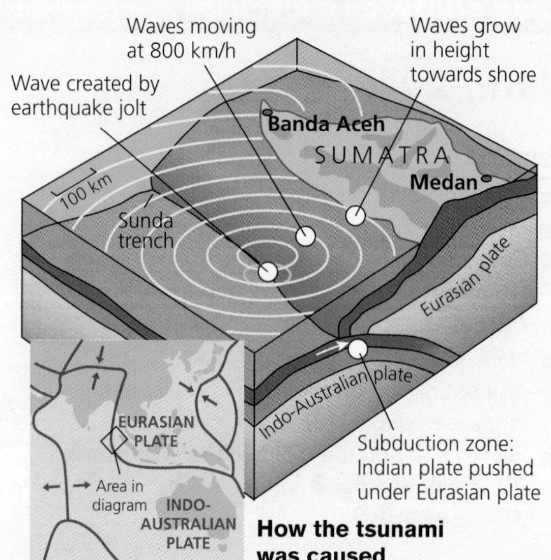

How the tsunami was caused

Effects

- Tidal wave swept across the Indian Ocean destroying coastal villages – there was complete devastation in some areas.
- Many people, including holidaymakers, did not recognise how destructive the wave would be and did not flee to higher ground and were drowned.
- Tidal wave destroyed coastal road, rail and telecommunications.
- Fishing and tourist industry destroyed.

Short-term response

- Relief teams were swamped by the scale of the disaster and many injured were not treated for days.
- Emergency food, fresh water, sheeting and tents poured in from around the world.
- Medical teams established aid stations to treat the injured and prevent disease from contaminated water.
- Tents and temporary shelters were erected to provide shelter for the homeless.
- Heavy equipment was brought to the area to clear roads destroyed by the force of the water.

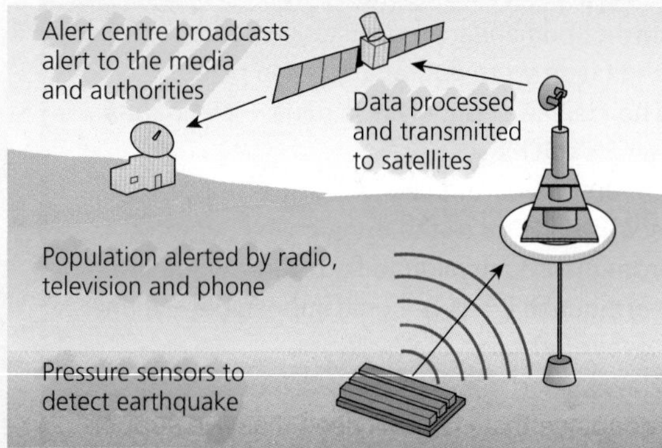

Tsunami early warning system

Long-term response

- The immense amount of money donated – £372 million by the British public alone – is being used to build new and stronger housing for the people affected.
- Rebuilding the fishing industry and coastal resorts for the tourist industry.
- Rebuilding the communications and power infrastructure of affected areas.
- By 2006, an early warning system like the one in the Pacific Ocean was in place, and education programmes ensure people know what to do in the event of a alert being sounded.

Tsunami alert – what to do

- Get off the beaches and move to higher ground.
- Stay away from rivers flowing into the ocean.
- Wait for the all clear – there may be more than one wave.

Test yourself

1 **Label the diagram correctly using words from the list.**

pyroclastic flow	lahar
volcano	hot spot
fold mountains	tsunami
caldera	mantle
new mountains	subduction zone
trench	plate

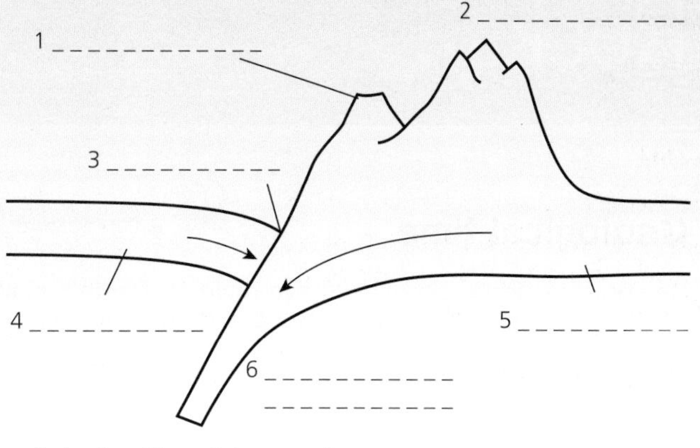

A destructive plate margin

2 **Match the correct word from the list to each description.**

- Hot gases, ash and steam which move very fast and can cause tremendous damage.
- A mudflow formed when hot ash melts snow and ice or falls into rivers.
- The very large depression in which a supervolcano will sit.
- Where light molten lava rises towards the surface and collects in a store beneath the surface.
- The deepest part of the oceans and formed at destructive plate margins.
- Formed where huge forces cause the rocks to buckle and are forced upwards.
- A tidal wave triggered by an earthquake.

Exam tip

Notice the command words 'describe' and 'explain'. Make sure you know what they are asking you to do – see 'Command words' on page 2. Make sure you write about a named example in question (a) to gain maximum marks.

Examination question

Foundation tier:

(a) For an earthquake you have studied:

(i) Name the earthquake.

(ii) Describe the effects of the earthquake. *(4 marks)*

(iii) Why are the effects of earthquakes more damaging in poorer countries than in richer countries. *(4 marks)*

Higher tier:

(b) Using examples from earthquakes you have studied, describe and explain why earthquakes in poorer countries are more often more destructive than in richer countries. *(8 marks)*

Rocks

Geological time

Geologists study rocks. They have divided the time since our major rocks were formed into **eras** (longer periods of time) and divided the eras into **periods**. The boundaries between the different time zones were decided by important times of mountain building or widespread sea level change.

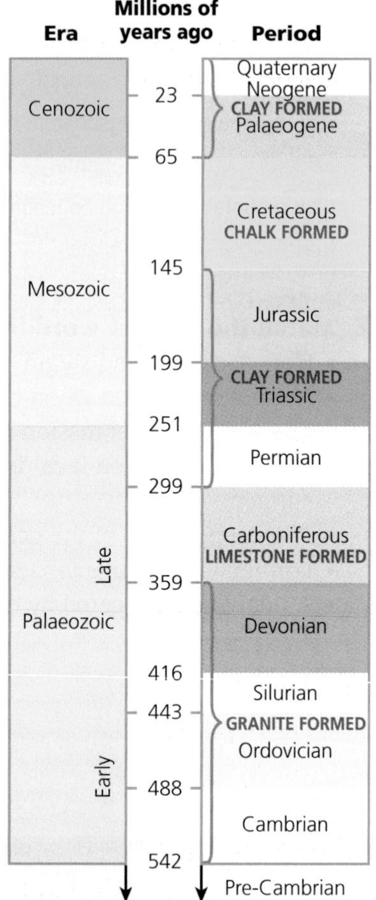

The geological timescale

Types of rocks

There are three main types of rocks:

Igneous rocks

These are formed when molten magma cools. If it cools beneath the surface, large crystals have time to grow and granite is formed. If it flows to the surface, quicker cooling results in basalt. They are hard, tough rocks and are resistant to erosion.

Sedimentary rocks

They are usually made of fine material deposited in lakes and seas. They form layers of rock called **beds** separated by **bedding planes**.

Sandstone is formed from sand compressed by the weight of other deposits on top of it. Chalk and limestone are formed from billions of shells of tiny sea creatures that fell to the sea bed when they died.

They are weaker and more likely to erode than igneous or metamorphic rocks.

Metamorphic rocks

They occur when igneous or sedimentary rocks are changed by great heat or pressure from volcanic eruptions or mountain building. Chalk and limestone are changed in this way to **marble**. They are hard, tough rocks and are resistant to erosion.

The rock cycle

The diagram shows how rocks are constantly being eroded and recycled.

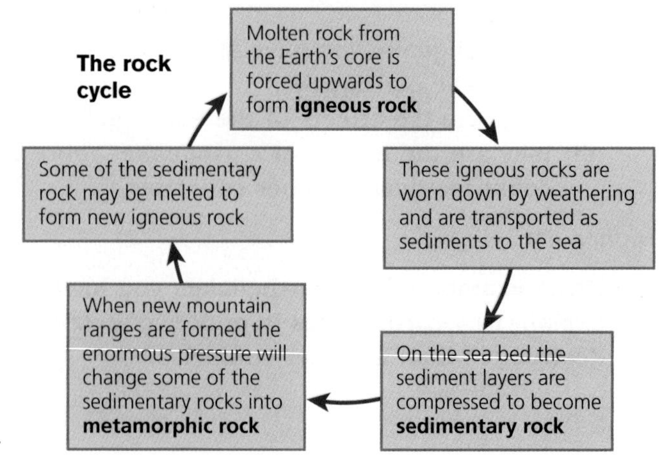

The rock cycle
Molten rock from the Earth's core is forced upwards to form **igneous rock**
These igneous rocks are worn down by weathering and are transported as sediments to the sea
On the sea bed the sediment layers are compressed to become **sedimentary rock**
When new mountain ranges are formed the enormous pressure will change some of the sedimentary rocks into **metamorphic rock**
Some of the sedimentary rock may be melted to form new igneous rock

Weathering

The rocks which form the Earth's surface are broken down by:

- **Erosion** by water in streams and rivers and by the sea; by ice in glaciers; by wind; and by people and animals (e.g. footpath erosion).
- **Weathering** by the constant attack of wind and rain and changes in temperature. In other words, the weather.

The whole process of breaking down rocks is called **denudation.**

Mechanical weathering

Mechanical weathering involves the breaking up of rocks by changes in temperature.

Cracks fill with water

Water freezes and expands as ice

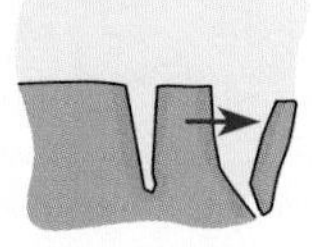

Cracks widen and pieces of rock split off

Freeze–thaw weathering

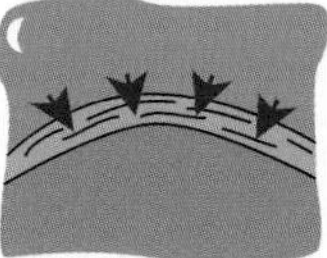

In hot areas heat of sun causes outer layer of rock to expand

Cold nights cause outer layer to contract

Outer layer flakes off

Exfoliation

Biological weathering

Biological weathering is caused by the action of plants or animals.

Biological weathering

Seed grows in a crack

Growing plant pushes against the rock and pieces loosen and break off

Chemical weathering

Chemical weathering is caused as some rocks dissolve in rainwater.

Solution Certain minerals and rocks dissolve in rainwater, e.g. rock salt, which is mined in Cheshire and used on roads in winter as a de-icer.

Carbonation

Rainwater becomes a weak carbonic acid

↓

This acidic rainwater reacts with the calcium carbonate of the limestone or chalk and forms calcium bicarbonate which is soluble

↓

As the carbonic acid gradually dissolves the calcium bicarbonate, the cracks in the rock are widened

↓

Limestone rock is formed from calcium carbonate ($CaCO_3$)

Key words

era	bedding planes	denudation	
period	metamorphic rock	mechanical weathering	chemical weathering
igneous rock	marble	freeze–thaw	solution
sedimentary rock	erosion	exfoliation	carbonation
beds	weathering	biological weathering	

Granite: landforms and landscapes

Granite areas of the British Isles

N

0 km 200

Granite

Granite is found in the British Isles in southwest England, the north of Scotland, northwest England and in Ireland.

Granite is an igneous rock, which is very resistant to erosion. It is formed from magma which has solidified beneath the surface. It is an **impermeable** rock, which means that water will not pass through it.

Granite consists of three main minerals:

Quartz – grey

Mica – black

Feldspar – pink, and will decay into a white clay when broken down by chemical weathering. This is known as **china clay** or **kaolin** and can be used to make delicate pottery

The formation of Dartmoor

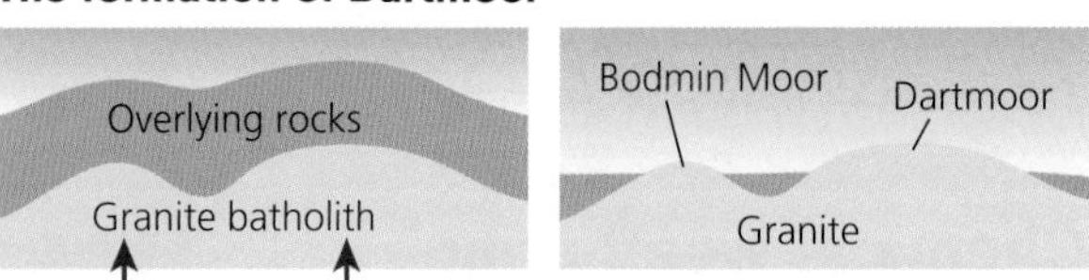

Magma forces its way upwards (intrudes) towards the surface, pushing up the overlying rocks. It cools beneath the surface to form granite. This mass of rock is called a **batholith**

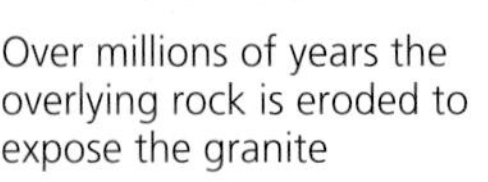

Over millions of years the overlying rock is eroded to expose the granite

Dartmoor granite

The moorland area of Dartmoor in southwest England is made of granite.

Granite tors on Dartmoor

The most distinctive features of the granite moorland landscapes are the **granite tors.** These are isolated **outcrops** or what look like mounds of rock on top of the hills.

Geologists suggest that the tors were formed when the overlying rocks were eroded during the ice age and the underlying granite became exposed. The areas of granite surrounding the tors had more cracks or **joints** and had been weathered more rapidly while underground than the areas of the tors which had fewer

A granite tor

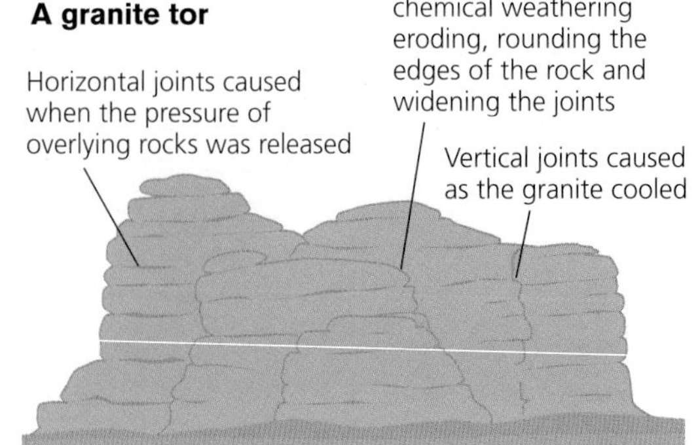

joints and so were more resistant. When the granite was exposed the tors were left standing higher than the surrounding land.

Tors are weathered into rounded shapes with numerous cracks by freeze–thaw action and some chemical weathering.

The granite landscape of Dartmoor

Granite is a strong rock resistant to erosion and is impermeable and produces a distinctively bleak landscape. In southwest England the granite upland of Dartmoor is:

- a wet and windswept upland area, with strong winds, heavy rain and low temperatures, often giving snow in winter
- mostly moorland with marshy bogs as the granite rock is impermeable with tough tussocky grasses and numerous streams and rivers which flow in steep sided valleys
- covered with generally thin and wet soils as the granite rock does not break down easily and only a few stunted trees are able to grow
- dotted with buildings made with the local granite which is a strong and waterproof local stone so fits naturally into the landscape

Farming on Dartmoor

The farm is a livestock farm keeping:

- sheep, usually Scottish Blackface kept on the moors in winter
- lambs sold in early autumn for fattening on lowland pastures
- cattle, usually hardy Galloways served by a Hereford bull producing good beef calves; after spending summer on the moor they are sold for fattening
- ponies which are kept on the moors and rounded up in the autumn; the foals are sold at pony markets

The high moorland is common land of heather and rough grassland – very exposed and cold in winter. Only sheep and Dartmoor ponies can survive in winter.

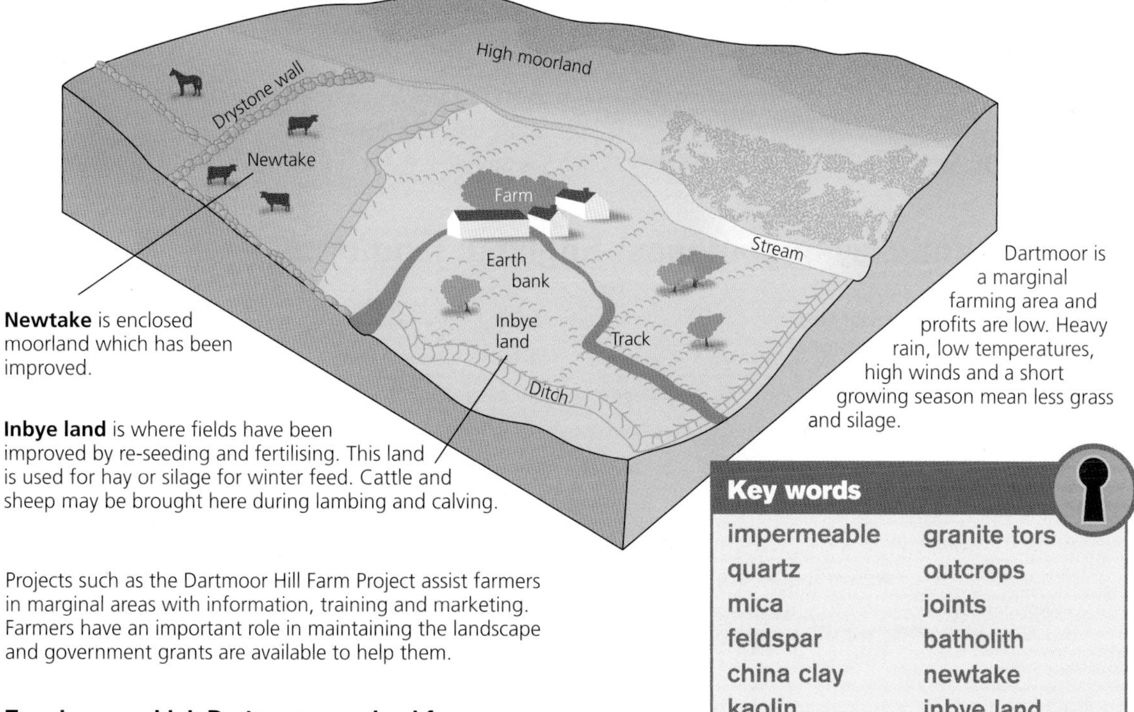

Newtake is enclosed moorland which has been improved.

Inbye land is where fields have been improved by re-seeding and fertilising. This land is used for hay or silage for winter feed. Cattle and sheep may be brought here during lambing and calving.

Dartmoor is a marginal farming area and profits are low. Heavy rain, low temperatures, high winds and a short growing season mean less grass and silage.

Projects such as the Dartmoor Hill Farm Project assist farmers in marginal areas with information, training and marketing. Farmers have an important role in maintaining the landscape and government grants are available to help them.

Farming on a high Dartmoor moorland farm

Key words

impermeable	granite tors
quartz	outcrops
mica	joints
feldspar	batholith
china clay	newtake
kaolin	inbye land

Chalk and clay: landforms and landscapes

Chalk and clay are mainly found close to each other in the south and east of England. The two rocks however are very different and give rise to different landscapes.

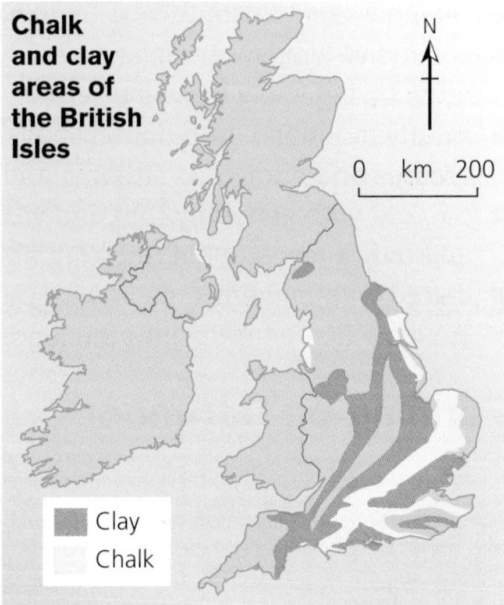

Chalk is:

- a sedimentary rock laid down in layers beneath the sea
- formed from the shells of billions of small sea creatures that lived in warm seas. They were crushed to form chalk which is a very pure type of calcium carbonate and vulnerable to chemical weathering
- quite a hard rock and it forms some high cliffs – the 'White Cliffs of Dover' are made of chalk – but it is not as resistant to erosion as granite
- a **permeable** rock, which means that water will soak into it through cracks and pores

Clay is:

- a sedimentary rock
- formed from rocks ground down into a flour by ice and deposited at the time of the last ice age
- a very soft rock and easily eroded by rivers so it forms areas of lowland
- an **impermeable** rock, which will not let water pass through it

Chalk and clay in southeast England: landforms

Landscape features of the chalk and clay of southeast England

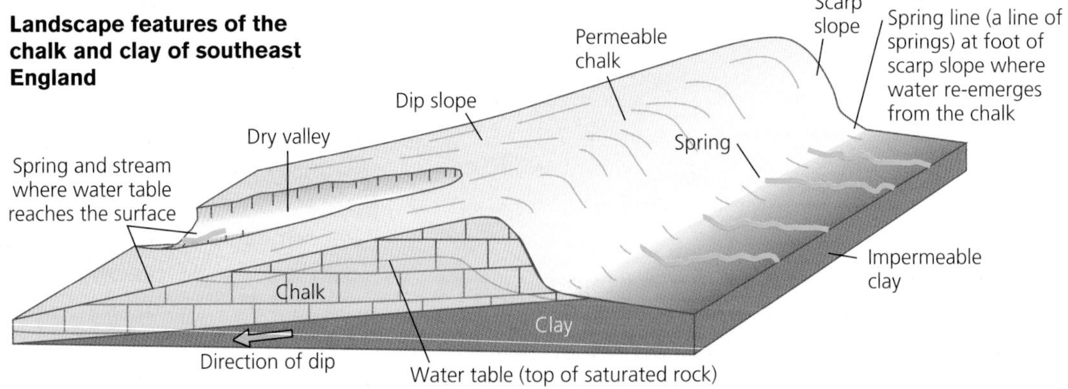

The dominant rocks of southeast England are chalk or clay.

Slopes

The chalk was laid down horizontally in warm shallow seas. When the Downs were formed they were gently pushed and tilted into a fold. The top of this fold was eroded leaving two sides, the North and South Downs. Each has a gentle or **dip** slope and a steep or **scarp** slope.

Springs

The chalk is porous so water soaks into it. Beneath the chalk is a layer of clay which is impermeable and will not let the water soak through it. The top of the water-saturated chalk is known as the **water table**. If you dig a well down to this level it will fill with water. At the foot of the chalk slopes, where the chalk meets the impermeable clay, the water seeps out of the rock in the form of springs.

Dry valleys

Dry valleys are valleys without a river or a stream and are common in chalk landscapes. They were formed by rivers which flowed during or following the ice age when the chalk was frozen and the water could not sink in, or when the ice melted and the chalk was so full of water it flowed over the surface.

Chalk and clay in southeast England: landscapes

Chalk landscapes have:

- rounded hills with gentle slopes and dry valleys
- poor thin soils as the water needed for plant growth soaks into the ground and takes all the nutrients with it
- short springy turf that in the past led to sheep farming but has now given way to growing crops
- large fields, divided up when farmers began to use fertilisers to enrich the poor soils and grow crops of wheat and barley

Clay landscapes have:

- flat land with streams and rivers as the rock is impermeable
- villages at the spring points at the foot of the chalk scarp slope
- deep soils which are heavy, sticky and waterlogged but hard and unworkable when dry
- wooded areas with fields with hedges growing grass and vegetable crops
- land used mainly for cattle farming

The chalk aquifer of the London Basin

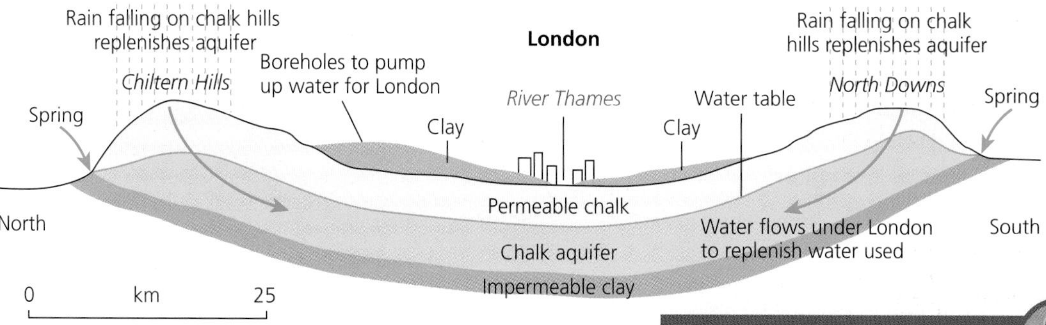

Beneath London the chalk rocks fold downward to form a **syncline**. The water soaks into the chalk on the North Downs and the Chilterns. It percolates down through the permeable chalk to collect beneath London in an **aquifer.** The water is pumped up to supply London.

Key words

permeable	water table
impermeable	dry valley
dip slope	syncline
scarp slope	aquifer

Carboniferous limestone: landforms and landscapes

Carboniferous limestone is mainly found in a band running down the centre of the northern half of England, particularly forming the Pennine hills. It is:

- a sedimentary rock, formed during the carboniferous period when billions of shells of small, dead sea creatures fell to the bottom of warm seas. Numerous fossilised shells of these creatures and corals are found in the rock
- a resistant and strong rock that has numerous vertical and horizontal cracks and forms steep rock **gorges** and cliffs
- a permeable rock that is composed of calcium carbonate and is very susceptible to chemical weathering. The weak acidic rainfall trickles down through the cracks and widens them

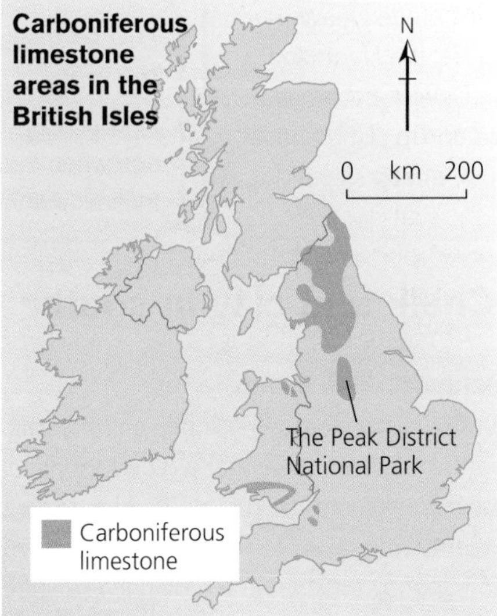

The landscape of the Peak District National Park

The landscape of Carboniferous limestone has:

- hills and steep-sided valleys. Some are dry valleys and others have rivers and streams where the underlying impermeable rocks are exposed
- thin soils as the water needed for plant growth runs down the cracks in the rock and takes all the nutrients with it
- rain and wind in the higher areas. Small fields are divided by drystone walls. In the sheltered valleys trees grow and the land is less harsh
- grazing cattle and sheep with little or no arable land

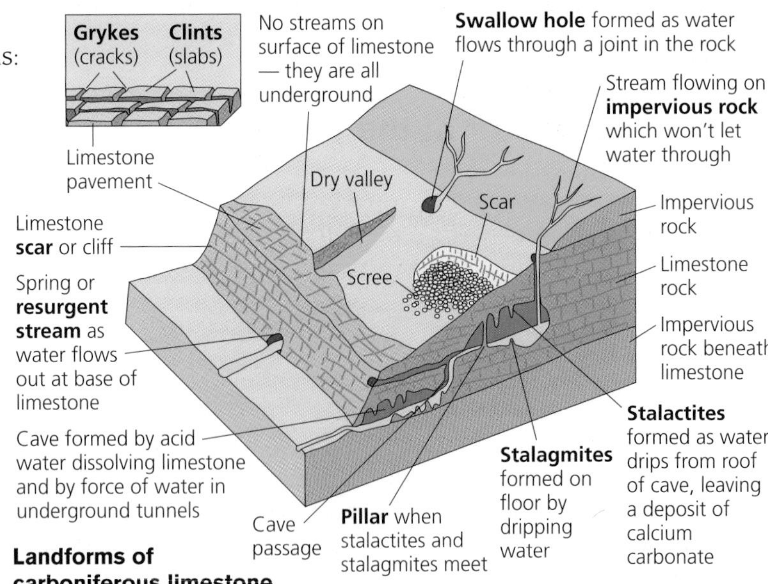

Landforms of carboniferous limestone

Tourism near Castleton in the Peak District National Park, Derbyshire

Opportunities for tourism

The Castleton area is very attractive with dramatic landscapes, spectacular views and pretty villages built of the local limestone, all conserved within a National Park that attracts an estimated 18–22 million tourists each year. There are good opportunities for walking and cycling on well-designated paths and trails, as well as caving, touring and climbing on the limestone cliffs.

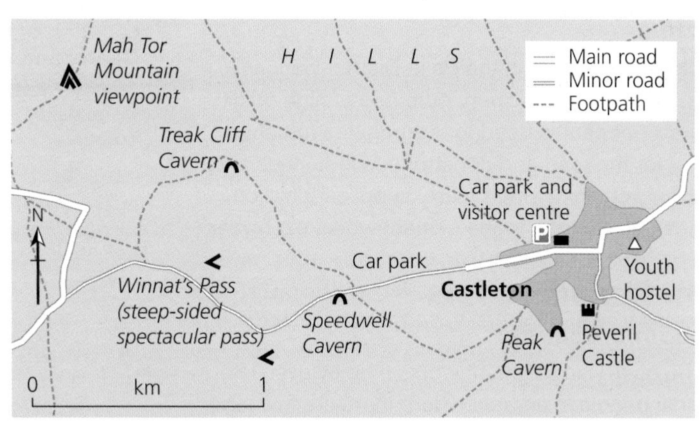

Location of Castleton

There are numerous facilities for tourists in Castleton and other surrounding villages and towns with shops, eating places, hotels, guest houses and campsites.

Benefits of tourism

Tourists bring money to the cafés, shops and hotels and so jobs are created for local people – tourism is the main employer in the National Park.

Local craftspeople – painters, woodworkers and jewellery makers – benefit, as do farmers supplementing their income by offering bed-and-breakfast or letting cottages. There are also guides, wardens and National Park officials who support the tourist trade.

Costs of tourism

- Heavy traffic on busy summer weekends and bank holidays trying to negotiate the narrow roads can create traffic jams and pollution.
- Castleton is one of the most attractive villages and becomes 'swamped' by the huge numbers of visiting tourists, with cars parked along all roads and local people unable to carry out their daily routines. Such villages are known as **honeypot villages** because they attract so many visitors.
- House prices to buy or rent in such an attractive area are higher than elsewhere and many local people are unable to afford their own home. Over 50% of houses in some popular villages are bought as second homes or as holiday lets leaving villages empty during the winter months.
- Shop and restaurant prices are often higher than elsewhere as owners strive to improve service to attract the tourist trade.
- Tourism brings with it difficulties. In the countryside, gates left open and walls knocked down allow animals to stray. Litter is dumped, which may harm livestock and wildlife.
- Tourism is a seasonal trade and many jobs are only available during the summer tourist season.

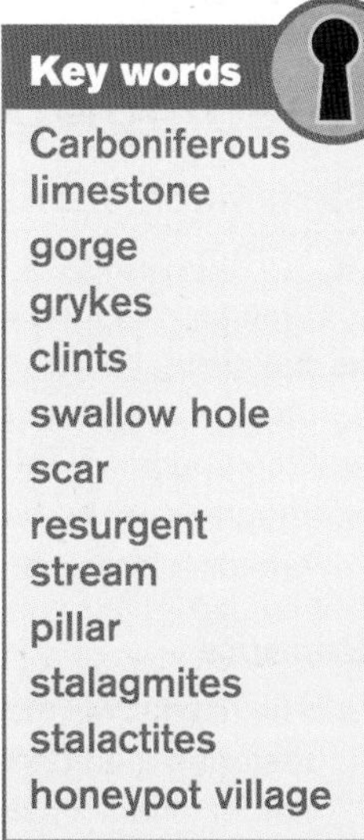

Case study: *a quarry*

Hope Quarry and cement works

Factfile

The Hope Quarry is in the Peak District National Park. The limestone is very pure and of high quality. Fifty-six per cent is used for aggregates (crushed stone) for the road and construction industries and 23% for cement. Hope Quarry supplies 2 million tonnes of limestone to the cement works each year.

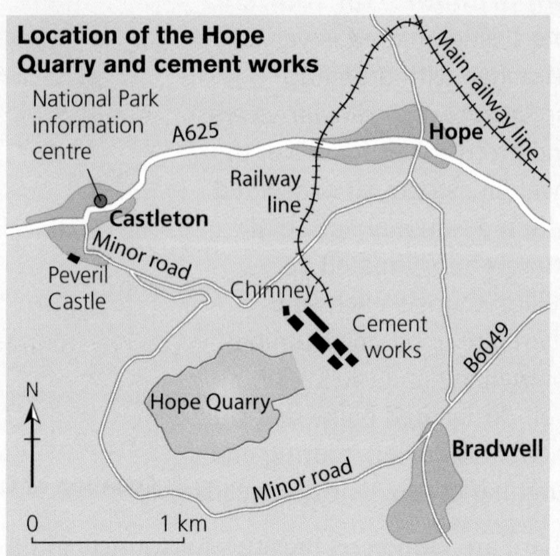

Benefits and costs

Economic benefits

- The quarry and cement works employ over 180 local people in an area where there are few jobs.
- These people use nearby shops and services so the whole area benefits.

Social benefits

- The quarry and cement works supply the aggregate and the cement which the UK needs to develop its roads and industries.

Environmental costs

- The cement works' buildings and tall chimney dominate the skyline in this beautiful valley.
- Dust from the quarry workings is unpleasant.
- Works' lorries are a hazard on the rural roads.

Reducing the environmental impact

- Landscaping to hide the quarry and works.
- Damping the dust with water sprays.
- Rail transport to reduce the impact of the lorries on local roads.
- 7,000 trees have been planted to offset the 1 million tonnes of carbon the cement works produces each year.
- Part of the area is a wetland wildlife reserve.

Case study: *quarrying and sustainable management*

Needingworth Quarry, Cambridgeshire

Sustainable management of the site

- In 1999, agreement was reached for a 700-hectare wetland reserve with footpaths, bridleways and cycleways.
- It is being built in stages as the quarrying work progresses and will include a reed bed with wet scrub and grassland.
- Reed beds will be cut on a 10-year rotation and the cut reeds sold.
- Progress will be checked by aerial photography and by surveying breeding birds, fish and plants every 5 years.

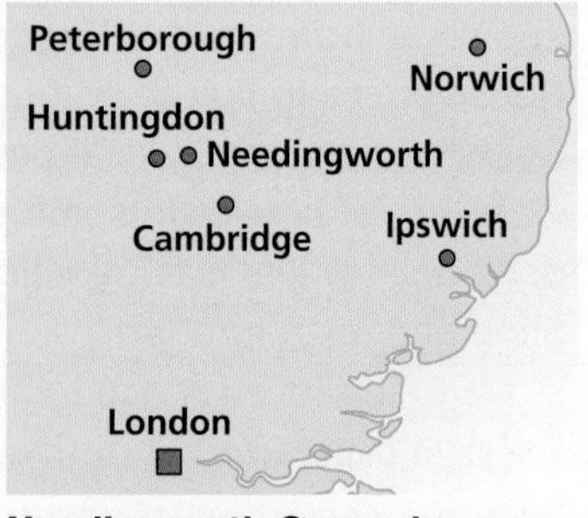

Needingworth Quarry is one of the largest sand and gravel extraction sites in the UK

Benefits

- The habitat will make a major contribution to the UK targets for restoring reed beds to attract rare birds like the bittern and marsh harrier, and animals like water voles and otters.
- The project will provide a major green space near the growing city of Cambridge.

GIS – Geographic Information System

GIS is a software package that allows storage, analysis and presentation of information about a location. The technology uses digital information from satellite and aerial imagery.

When a quarry site is being restored, GIS enables the geology, soil type, contours and height of the water table to be overlaid on an aerial photograph. This would accurately inform decisions about the creation of the most suitable type of vegetation and habitat.

Test yourself

1 **Match the following types of rock terms to the correct definition and example in the table.**

sedimentary rocks   igneous rocks   metamorphic rocks

Type of rock	Definition	Example
	Are usually made of fine material deposited in lakes and seas	Chalk
	Occur when rocks are changed by great heat or pressure	Marble
	Are formed when molten magma cools	Granite

2 **What is the difference between:**
- **(i) erosion and weathering?**
- **(ii) biological and chemical weathering?**

3 **Draw a diagram to show 'The rock cycle'.**

Exam tip

If you are taking the higher tier paper make sure you answer all the parts of a question such as (b)(ii) below.

Examination question

Foundation tier:

(a) For the scenery of a rock type you have studied:
- **(i) Name the rock type and an area where it is found.** *(1 mark)*
- **(ii) Describe *two* opportunities for tourism in the area you have chosen.** *(2 marks)*
- **(iii) Give *two* advantages of tourism in the area you have chosen.** *(4 marks)*
- **(iv) Give *two* disadvantages of tourism in the area you have chosen.** *(4 marks)*

Higher tier:

(b) For the scenery of a rock type you have studied:
- **(i) Name the rock type and the area where it is found.** *(1 mark)*
- **(ii) Describe the main opportunities for tourism in the area and the advantages and disadvantages of tourism in the area.** *(10 marks)*

The UK climate

The difference between weather and climate

Weather means day-to-day changes in atmospheric conditions. Words like sunny, warm, cloudy, rainy, windy or snowy are ways of describing the weather.

Climate is the pattern of weather over a year. It is put together by taking the average conditions over a period of 40 years. When describing the climate of an area you can include:

- the pattern of precipitation during the year (this might be snow or sleet as well as rain)
- the total precipitation for the year
- the average monthly temperature
- the range of temperature (the difference between the highest and lowest)
- average sunshine hours per month; average wind speeds and direction

Climate is often shown using a climate graph such as the one for Nottingham, England

Rainfall (mm) 200 150 100 50 0
Range of temperature 13°C
Total annual rainfall 810 mm
Temperature (°C) 20 15 10 5 0
J F M A M J J A S O N D
Months

Winters are mild with an average 3°C in January.

Rain in Britain falls throughout the year with a slight winter maximum.

Total rainfall in Nottingham averages 810 mm a year.

Summer temperatures are warm, reaching an average of 16°C.

Climate graph for Nottingham

Why it rains

The UK is much wetter in the north and west than in the south and east, and high land has much more rain than low-lying land. Rainfall in the Scottish Highlands is over 1,500 mm a year whereas in London it is only about 600 mm a year.

Why it rains

→ Warm air → Cool air

Relief rainfall

Relief (orographic) rain helps to explain why north and west Britain are much wetter than the south and east. The **prevailing wind** in Britain is from the southwest and this brings moist air which cools as it is forced to rise over the Pennines, Welsh mountains and Scottish Highlands. Condensation causes clouds and rain.

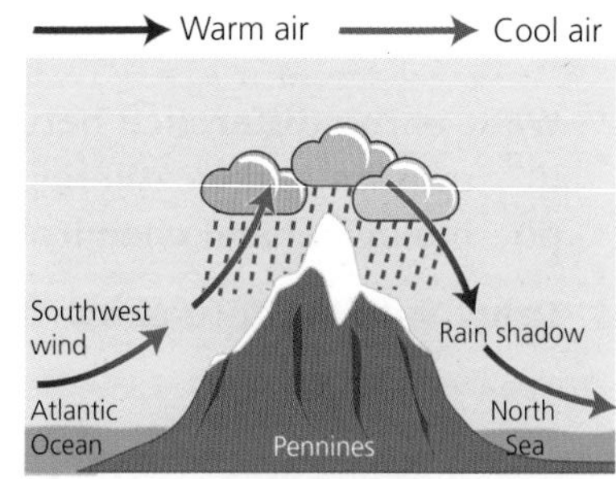

Frontal rainfall

Frontal rain develops along weather fronts which form where two air masses meet. Warm moist air meeting cooler air at a warm front is forced upwards. As it rises it cools and the moisture condenses to form rain. Britain lies in an area where air masses frequently meet and so we get a lot of this type of rain.

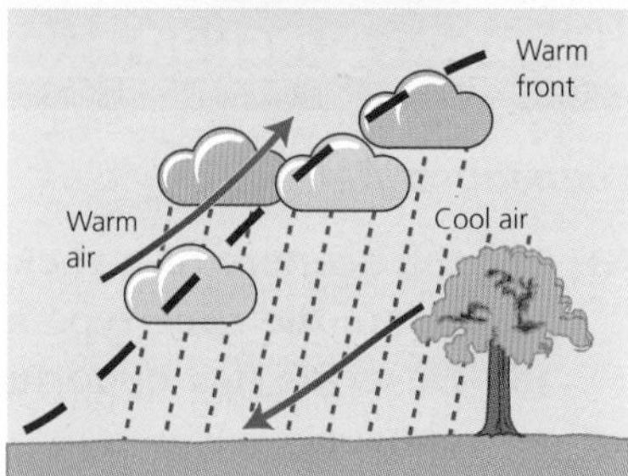

Convectional rainfall

On sunny days air warms and rises. As it rises it cools and the water vapour in the air condenses to form clouds and rain. Convectional rainfall occurs less than the other two types of rain in Britain because it is caused by heat, but long hot summer days can lead to afternoon thunderstorms, especially in southeast England.

Reasons for the variation in the UK climate

The UK has a temperate maritime climate. **Temperate** because it is neither as hot as the tropics nor as cold as the poles. **Maritime** because it is greatly influenced by the sea.

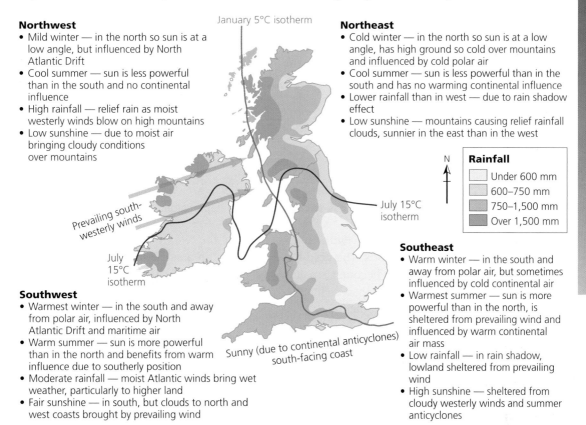

Northwest

- Mild winter — in the north so sun is at a low angle, but influenced by North Atlantic Drift
- Cool summer — sun is less powerful than in the south and no continental influence
- High rainfall — relief rain as moist westerly winds blow on high mountains
- Low sunshine — due to moist air bringing cloudy conditions over mountains

Northeast

- Cold winter — in the north so sun is at a low angle, has high ground so cold over mountains and influenced by cold polar air
- Cool summer — sun is less powerful than in the south and has no warming continental influence
- Lower rainfall than in west — due to rain shadow effect
- Low sunshine — mountains causing relief rainfall clouds, sunnier in the east than in the west

January 5°C isotherm

Prevailing south-westerly winds

July 15°C isotherm

July 15°C isotherm

Sunny (due to continental anticyclones) south-facing coast

N

Rainfall

- Under 600 mm
- 600–750 mm
- 750–1,500 mm
- Over 1,500 mm

Southwest

- Warmest winter — in the south and away from polar air, influenced by North Atlantic Drift and maritime air
- Warm summer — sun is more powerful than in the north and benefits from warm influence due to southerly position
- Moderate rainfall — moist Atlantic winds bring wet weather, particularly to higher land
- Fair sunshine — in south, but clouds to north and west coasts brought by prevailing wind

Southeast

- Warm winter — in the south and away from polar air, but sometimes influenced by cold continental air
- Warmest summer — sun is more powerful than in the north, is sheltered from prevailing wind and influenced by warm continental air mass
- Low rainfall — in rain shadow, lowland sheltered from prevailing wind
- High sunshine — sheltered from cloudy westerly winds and summer anticyclones

Characteristics of the UK climate

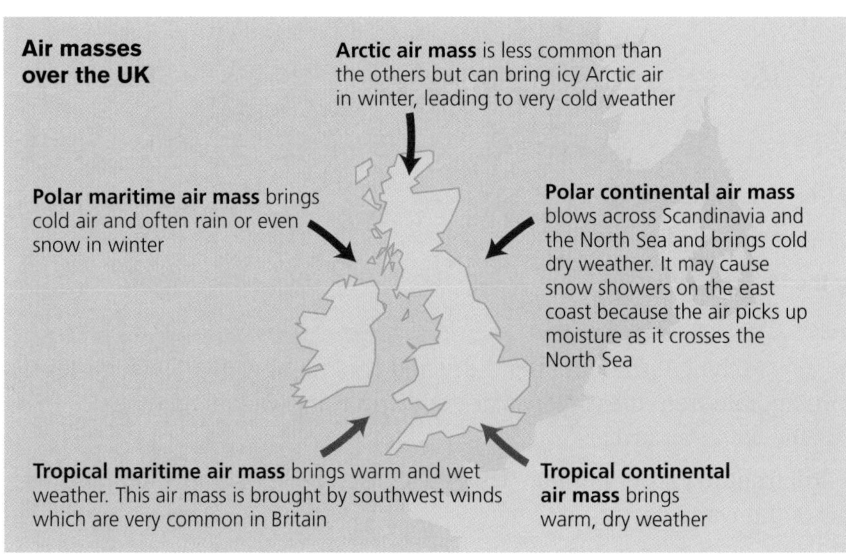

Air masses over the UK
Arctic air mass is less common than the others but can bring icy Arctic air in winter, leading to very cold weather
Polar maritime air mass brings cold air and often rain or even snow in winter
Polar continental air mass blows across Scandinavia and the North Sea and brings cold dry weather. It may cause snow showers on the east coast because the air picks up moisture as it crosses the North Sea
Tropical maritime air mass brings warm and wet weather. This air mass is brought by southwest winds which are very common in Britain
Tropical continental air mass brings warm, dry weather

Key words

- weather
- climate
- prevailing wind
- temperate
- maritime
- air mass

Depressions and anticyclones

Depressions

Depressions are low-pressure systems that bring changeable, usually rainy weather. They often develop to the west of Britain, where tropical and polar air masses meet, and travel eastwards across the country, blown by the prevailing westerly winds.

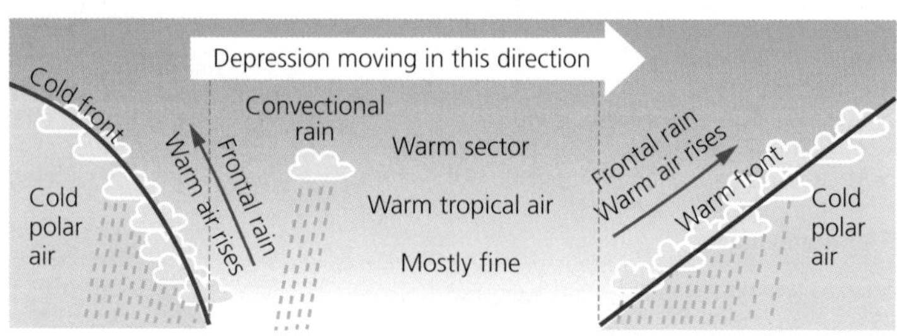

A cross-section through a depression

A depression is recognisable on a weather map because of the two fronts, a warm front and a cold front, which meet at an apex. The **cold front** moves more quickly than the **warm front** so the **warm sector** gradually becomes smaller. As the fronts merge an **occluded front** forms, often bringing particularly bad weather conditions. As a depression crosses over the country it brings changeable weather.

Forecast for Birmingham

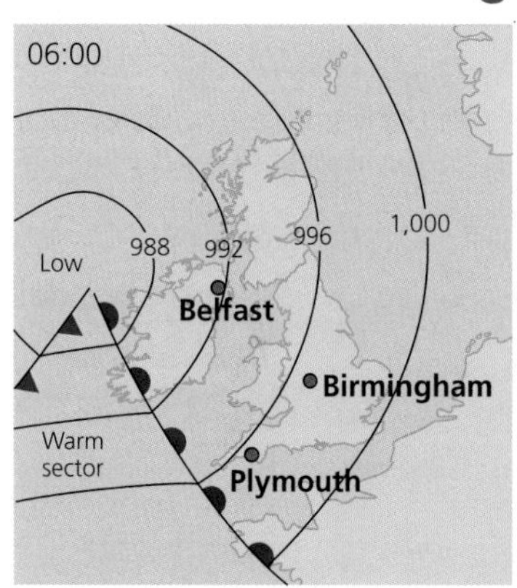

The depression is approaching Ireland

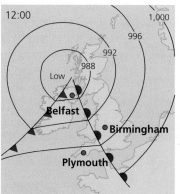

The depression has moved northeast and is now centred over western Scotland

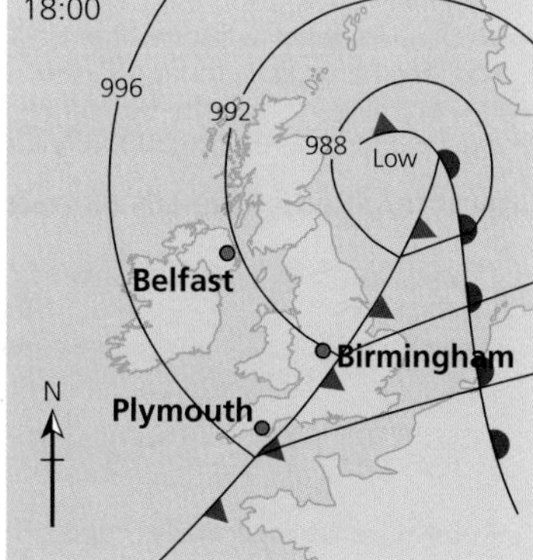

The depression is moving away from Britain. The cold front is now lying across southeast England

A depression crossing the UK

From the map above:

06.00 – clouds will be approaching Plymouth after a dry and cool night with a few high-level clouds. During the morning, rain from these clouds at the warm front will fall and the temperature will rise as the warm air arrives

12.00 – behind the warm front the air is warmer. There will be a spell of sunny intervals and showers in the mild air of the warm sector

18.00 – the cold front will bring rain. By 18.00, as the cold front passes, cooler air will return

Anticyclones

Anticyclones are high-pressure systems. These large areas of descending air bring settled weather that can last for several days or weeks. Pressure in these systems is usually over 1,000 millibars. If they persist they are called 'blocking highs' because they block out depressions.

In winter:

- Anticyclones bring cold, usually clear, settled weather. Rotating clockwise they drag cold Arctic air down from the north in an Arctic air mass or a polar continental air mass (see map, right).
- Temperatures can remain well below freezing, although the days may be clear and sunny with few clouds.
- At night it gets even colder because the lack of cloud cover means heat is lost quickly from the ground. This weather can cause problems, especially icy roads.
- Sometimes winter anticyclones contain a great deal of moisture in a polar maritime air mass (see map, page 29) and this produces dull, cloudy conditions that can last for many days. **Fog** may develop in the early mornings.

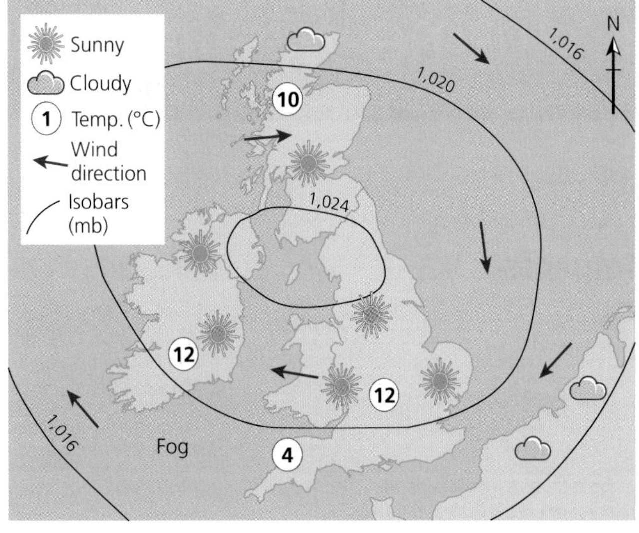

There are no weather fronts and this means that high-pressure systems generally bring dry weather

The **isobars** are circular and spaced far apart, so the winds tend to be light and blow out from the centre in a clockwise direction

An anticyclone

In summer:

- Anticyclones bring hot, sunny days with little or no cloud with temperatures up to 25°C or higher.
- At night, temperatures fall rapidly because there is no cloud cover and this can cause heavy morning **dew**.
- Thunderstorms may occur in the late afternoon of a hot day. These are caused by convection (see diagram, page 28) and may be violent.
- In the centre of the anticyclone the skies will be clear. Towards the edges of the anticyclone where the conditions are less settled there will be some cloud. Where winds reach the coastal area across the sea, e.g. the North Sea coasts, they may also experience more cloud.
- Summer anticyclones can cause **heatwaves** and drought. As the vegetation dries out, **moorland and forest fires** may break out. Sunshine can cause a build-up of pollutants in the atmosphere, especially in towns, and the lack of wind in an anticyclone means that they do not disperse.

Key words

depression	isobar
cold front	dew
warm front	heatwave
warm sector	drought
occluded front	moorland and forest fires
anticyclone	
fog	

Extreme weather in the UK

Extreme weather differs significantly from the average or what is expected.

Impacts

Weather	Impact
Drier than average years: 2005	Water shortages in the southeast of the country leads to restrictions on its use. Water Boards have built grid schemes for moving water around the country
Colder than expected: winter 2009/2010	A period from December to March with temperatures below freezing and above average snowfalls. Heating costs rose. Roads treacherous
Higher than average summer temperatures: 2003	Those with lung and heart conditions can find it difficult; danger of skin cancers
Higher than average rainfall: 2007/8/9	Sudden outbursts of exceptionally heavy rainfall cause flooding, damaging homes, businesses and farms

Case study: *the impact of an extreme weather event*

Flooding in Cumbria, 2009

Factfile

- Severe flooding in Cockermouth and Workington as the River Derwent and River Cocker overflowed their banks.
- The town centre of Cockermouth under 2.5 metres of water affecting 160 properties.

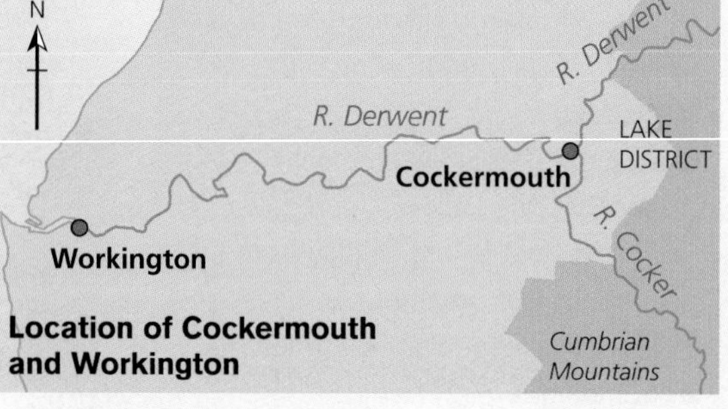

Location of Cockermouth and Workington

Cause of flooding

- In 24 hours, an unprecedented 320 mm (8+ inches) of rain fell on the hills around Seathwaite in the Lake District.
- The ground was quickly saturated and the water ran into the rivers and surged towards the towns of Cockermouth and Workington.

The effects

- Homes and shops were flooded and people were stranded.
- Two bridges collapsed. Roads were blocked, schools closed, power lines brought down.
- Fields alongside the rivers were flooded and covered with rock and debris.

Relief efforts

- Rescue helicopters, the RNLI and emergency services rescued people from their homes, emergency plans were in place and worked well.
- Over 200 people had to take refuge in emergency accommodation.

The long-term response

- Householders and business owners cleared up.
- The Army built a temporary bridge to link both sides of the town of Workington. A temporary rail station was also built.
- Seven bridges are to be rebuilt, roads repaired and farmers received grants to clear debris from their fields.
- A 0.75 km (1/2 mile) long bank was built alongside the River Derwent to protect the people of Cockermouth.
- Agencies meet to manage the response in case this ever happens again.

Global climate change

The overwhelming conclusion by the UN Intergovernmental Panel on Climate Change (IPCC) is that **global climate change**, particularly global warming, is taking place.

Evidence for global climate change

- Average global temperatures over the last 150 years show an upward trend.
- Global rise in temperature from 1976 to 2003 was three times greater than in the whole of the previous 100 years.
- The highest average global temperatures on record all occurred between 1998 and 2010.
- The area of the Arctic ice cap which does not melt in summer has shrunk by 9% since 1978.
- Glaciers around the world are retreating and sea levels are rising.

Evidence against global climate change

- The global warming that is taking place is part of a natural cycle.
- We are living in a post-glacial age. The last glacial period ended around 10,000 years ago and the temperatures have continued to rise since.
- Even in historical time there have been fluctuations in temperatures.

The causes of global climate change

The increase in carbon levels released into the atmosphere is so closely mirrored by the increase in global temperatures that scientists agree that the increase in global temperatures has been caused by the activities of people, especially in industrialised countries.

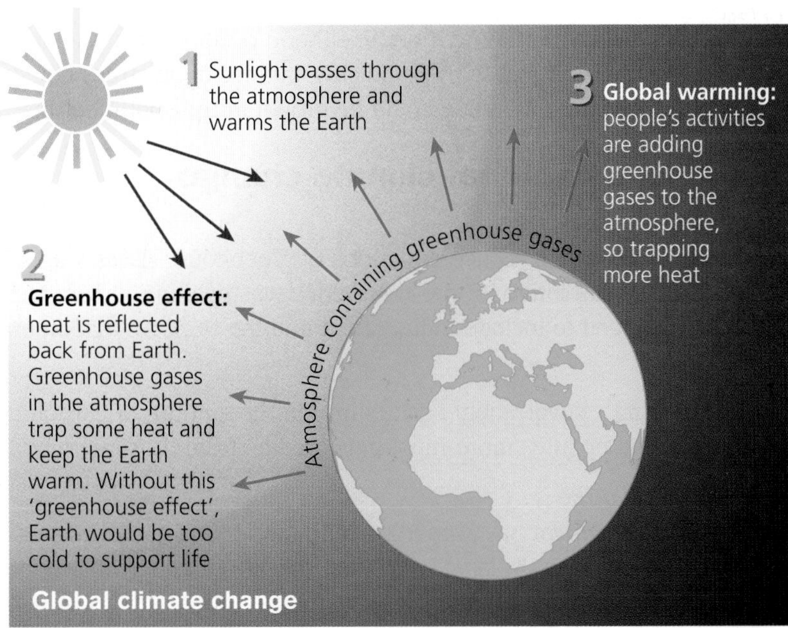

1 Sunlight passes through the atmosphere and warms the Earth
2 Greenhouse effect: heat is reflected back from Earth. Greenhouse gases in the atmosphere trap some heat and keep the Earth warm. Without this 'greenhouse effect', Earth would be too cold to support life
3 Global warming: people's activities are adding greenhouse gases to the atmosphere, so trapping more heat
Atmosphere containing greenhouse gases

Global climate change

Warming is caused by an increase in:

- carbon dioxide and oxides of nitrogen released by burning fossil fuels
- methane released by breakdown of waste
- chlorofluorocarbons from fridges and aerosols (now banned)

The effects of global climate change

Wildlife will find it hard to adapt to changes in climate and habitat. Polar bears, which hunt seals on the sea ice are already finding it difficult to adapt to the thawing of the Arctic ice

Climatic belts will expand north and south. Expansion of drought-prone areas to the north and south of desert regions could affect water supplies and agriculture in southwestern USA, the Mediterranean, southern Australia, southern Africa and parts of South America

Sea levels are rising by 0.2 cm per year. The ice cap in the Arctic has decreased by 15% since 1960 and the ice has thinned by 40%

Patterns of rainfall will change. Areas with plenty of rainfall will receive more and regions with low rainfall will receive less, affecting crop growth

The effects of global climate change

How will global climate change affect Britain?

- Rising sea levels of 20–40 cm mean low-lying areas will be flooded.
- More gales and storms in winter; warmer, drier summers.
- Britain could become a more popular holiday destination.
- There will be more rain and more likelihood of flooding in the autumn and winter.
- Coniferous trees will grow more quickly, but deciduous trees could die.
- Vines, sunflowers and maize could be grown in the south, but insect pests could flourish.

Responses to global climate change

International response

The **Kyoto Protocol** to reduce emissions of greenhouse gases was signed by many countries in 1997. In 2010, an international conference was held in Cancún, Mexico to strengthen the Kyoto Protocol and to include India, China and the USA. An agreement was reached but no binding treaty.

Carbon trading – governments set a limit on the amount of carbon a company can emit. A company that pollutes more must buy 'credits' from those who pollute less.

National response

The British government has agreed to cut emissions of greenhouse gases by:

- setting targets – the British government is committed to reduce emissions by 60% by 2050
- using natural gas or renewable energy instead of coal/oil to generate electricity
- using new technology such as **carbon capture**
- improving insulation in public buildings such as schools
- persuading people to save energy in their homes
- local authorities must reduce emissions from transport and landfill

Key words

Kyoto Protocol
carbon trading
carbon capture

Tropical revolving storms

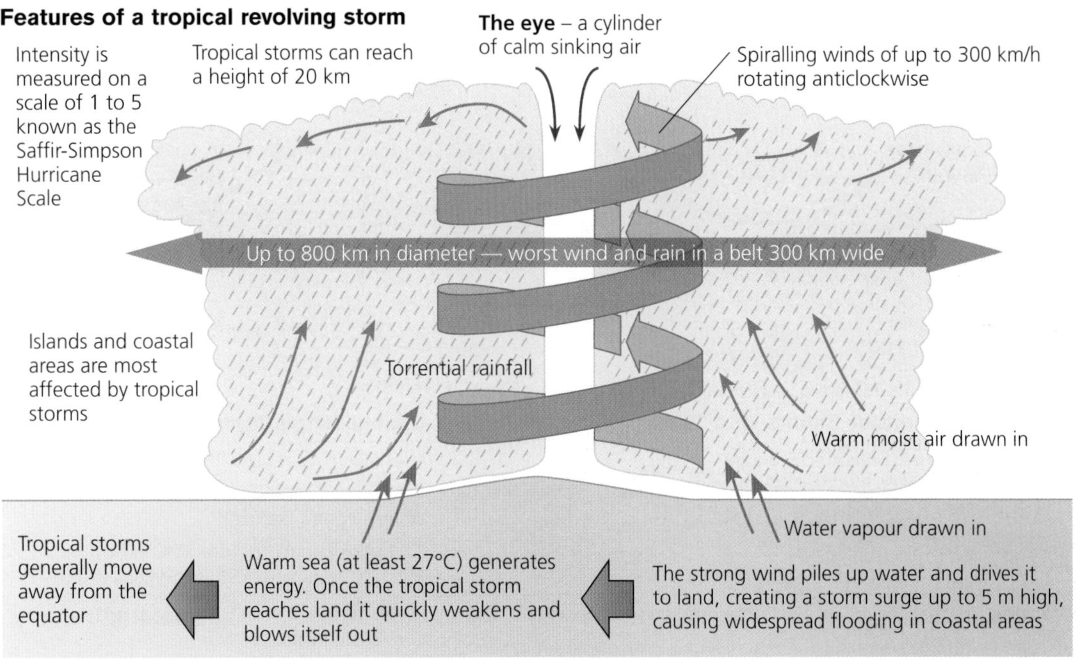

Intensity is measured on a scale of 1 to 5 known as the Saffir-Simpson Hurricane Scale

The eye – a cylinder of calm sinking air
Spiralling winds of up to 300 km/h rotating anticlockwise
Up to 800 km in diameter — worst wind and rain in a belt 300 km wide
Islands and coastal areas are most affected by tropical storms

Warm moist air drawn in
Water vapour drawn in
Tropical storms generally move away from the equator
Warm sea (at least 27°C) generates energy. Once the tropical storm reaches land it quickly weakens and blows itself out
The strong wind piles up water and drives it to land, creating a storm surge up to 5 m high, causing widespread flooding in coastal areas

Impacts

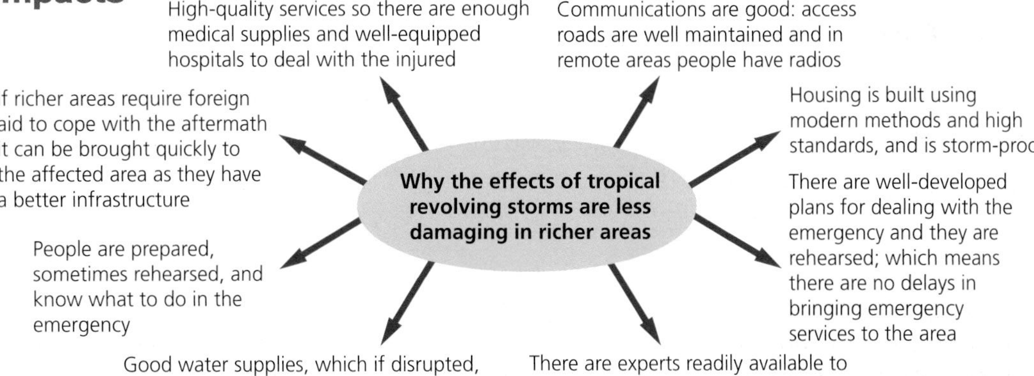

Why the effects of tropical revolving storms are less damaging in richer areas
High-quality services so there are enough medical supplies and well-equipped hospitals to deal with the injured
Communications are good: access roads are well maintained and in remote areas people have radios
If richer areas require foreign aid to cope with the aftermath it can be brought quickly to the affected area as they have a better infrastructure
Housing is built using modern methods and high standards, and is storm-proof
There are well-developed plans for dealing with the emergency and they are rehearsed; which means there are no delays in bringing emergency services to the area
People are prepared, sometimes rehearsed, and know what to do in the emergency
Good water supplies, which if disrupted, can be quickly repaired. This stops the spread of water- borne disease in the aftermath of the storm
There are experts readily available to give advice and assess the situation

Key words

eye of the storm

tropical revolving storm

Case study: comparing two tropical revolving storms

Tropical storms in a rich and a poor country

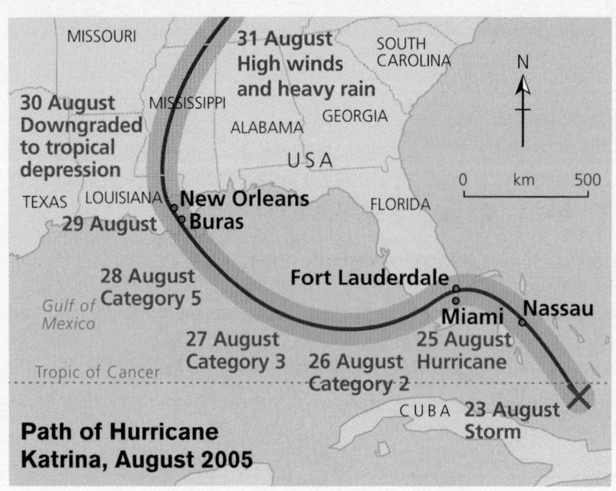

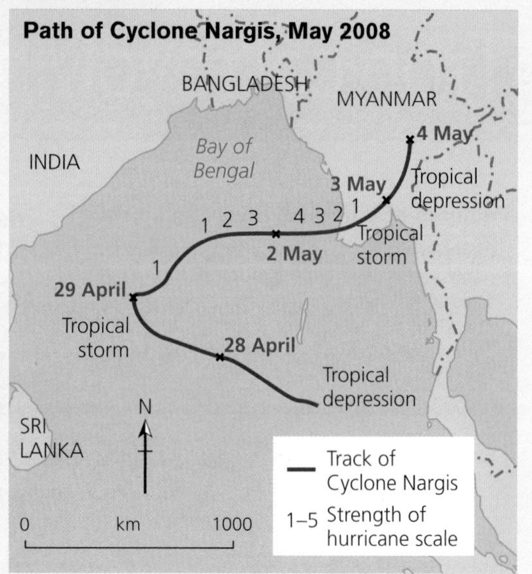

	Hurricane Katrina, USA 2005	Cyclone Nargis, Myanmar 2008
The facts	The USA is a rich country Highest wind speed 280 km/h Category 5 storm Damage – 1,836 people confirmed dead, 705 missing Costliest hurricane in history – est. $200 billion	Myanmar is a poor country Wind speeds of 217 km/h with torrential rain and storm surges 7.6 m high 1.5 million people severely affected, over 125,000 dead and 95% of buildings destroyed
Warning	National Hurricane Center warned of its approach 28th August – 1.2 million people issued with evacuation order, motorways jammed as people left the city, emergency centres set up	Storm predicted to hit Bangladesh and suddenly turned into Myanmar Most people knew nothing of the cyclone's approach
Effects	High winds – trees fall blocking roads, windows blown out, damage to high buildings Heavy rainfall with 200–250 mm, maximum 380 mm, contributed to flooding Storm surge in excess of 4.3 m. Flooding across 80% of the city with loss of life. Bridges washed out, electricity, gas and water services destroyed	High winds ripped roofs off buildings and destroyed houses Heavy rainfall – sewage systems unable to cope, fear of disease Storm surge made people homeless, many drowned, rice crops in the Irrawaddy Delta destroyed by flooding
Short-term response	Difficulties in communications, people stranded without food and clean water, possibility of the spread of disease Emergency housing The coastguard rescued over 33,500 people 58,000 National Guard and police mobilised to help with law and order, rescue and clean up	Foreign aid workers faced restrictions Reported that foreign aid was restricted to the cities and little reached the countryside Workers provided medicines, safe water and food but aid was disorganised
Long-term response	The US Congress authorised $62.3 billion in aid Victims provided with temporary accommodation Plans made to strengthen flood defences	Little is known of the response Rebuilding transport networks, services, homes and factories Reclaiming agricultural land

Test yourself

1 What is the difference between weather and climate?

2 Draw diagrams to show how the three types of rainfall are formed (relief, convectional and frontal).

3 Name three of the main air masses affecting Britain and describe the typical weather associated with each.

4 What is meant by the term 'prevailing wind'?

5 Explain why the climate of Plymouth is different from that of Aberdeen.

6 Draw a diagram to show the features of a tropical revolving storm.

7 Give two of the conditions needed for a tropical revolving storm to develop.

8 Give four reasons that show the impact of a tropical revolving storm is greater in a poorer country than in a richer country.

Exam tip

Even 'short' questions carry 2 marks. Always make sure you have two main points in your answer. In some questions you are asked specifically for two, three or four points. Make sure you include the required number to gain marks.

Examination question

Foundation tier:

(a) (i) Name three ways in which people are adding greenhouse gases to the atmosphere. *(3 marks)*

(ii) Describe four effects global warming would have on Britain. *(4 marks)*

(iii) Give ways in which individuals and local authorities can help to reduce carbon emissions in their area. *(6 marks)*

Higher tier:

(b) What is the evidence for climate change and that it is caused by 'greenhouse gases' such as carbon dioxide? *(3 marks)*

(c) How will climate change affect Britain? *(4 marks)*

(d) How are the British government hoping to combat climate change at an international and national level? *(6 marks)*

Ecosystems

An **ecosystem** consists of the plants and animals that live together in a particular environment. They depend on each other and on non-living parts of the environment such as rock and soils, water, sunlight and temperature.

The structure of a woodland ecosystem

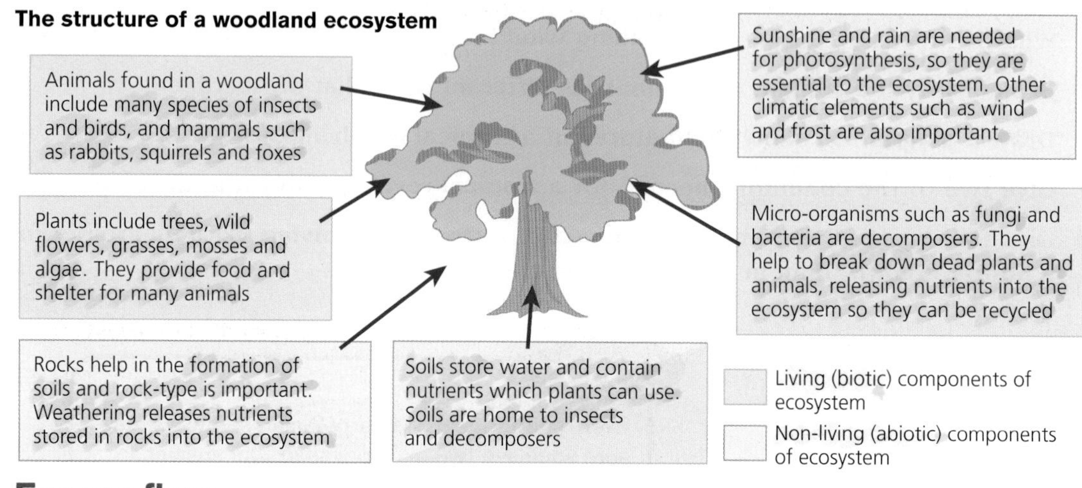

Animals found in a woodland include many species of insects and birds, and mammals such as rabbits, squirrels and foxes
Plants include trees, wild flowers, grasses, mosses and algae. They provide food and shelter for many animals
Rocks help in the formation of soils and rock-type is important. Weathering releases nutrients stored in rocks into the ecosystem
Soils store water and contain nutrients which plants can use. Soils are home to insects and decomposers
Sunshine and rain are needed for photosynthesis, so they are essential to the ecosystem. Other climatic elements such as wind and frost are also important
Micro-organisms such as fungi and bacteria are decomposers. They help to break down dead plants and animals, releasing nutrients into the ecosystem so they can be recycled
Living (biotic) components of ecosystem
Non-living (abiotic) components of ecosystem

Energy flow

All energy comes from the sun. Green plants use sunlight to make their own food (glucose) through the process of **photosynthesis**. Water and carbon dioxide are also essential for photosynthesis to occur. Green plants are known as **primary producers** because of their ability to produce food.

How do ecosystems function?

All ecosystems have two processes:

1 The flow of energy through the system.
2 The cycling of nutrients.

If either of these two processes is disrupted, the whole ecosystem will be affected.

Nutrient cycling

Nutrients are essential minerals such as nitrogen, calcium and sodium which are needed by all living things. Nutrients occur naturally in rocks, water and the atmosphere and can enter an ecosystem from any of these **nutrient pools.**

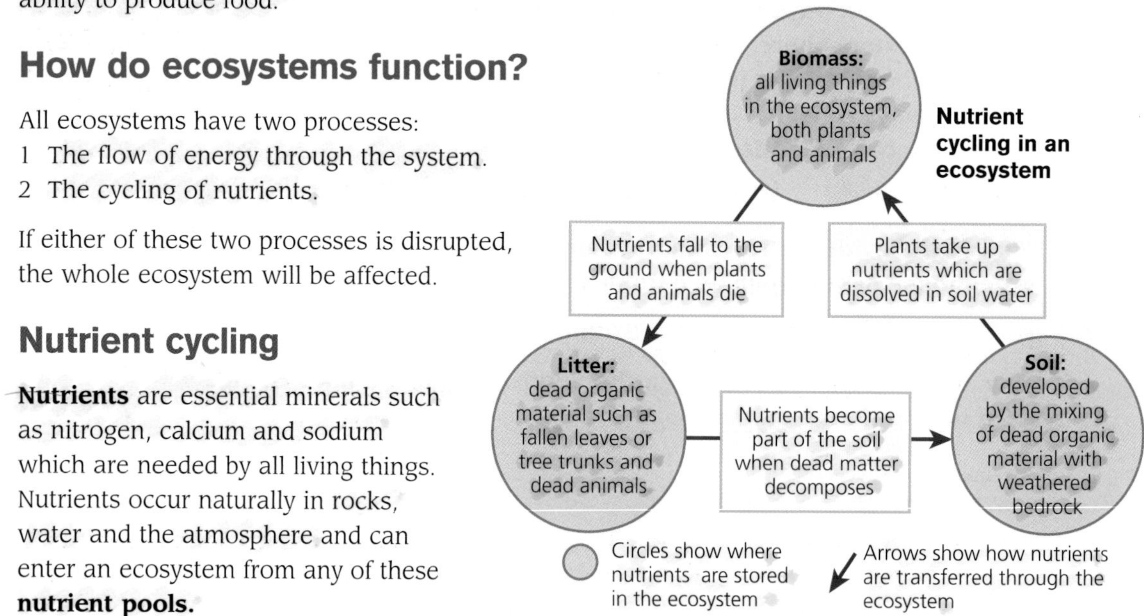

Plants take up nutrients which are dissolved in soil water
Litter: dead organic material such as fallen leaves or tree trunks and dead animals
Nutrients become part of the soil when dead matter decomposes
Soil: developed by the mixing of dead organic material with weathered bedrock

Nutrients may also be lost from an ecosystem. They can, for example, be washed away in surface runoff or by **leaching**. Leaching occurs in areas where there is plenty of rain and good soil drainage. Soluble salts such as calcium and magnesium are washed downwards through the soil and re-deposited in the lower layers out of reach of plant roots. The upper layers become increasingly acid and less fertile.

Food webs and chains

Energy is passed on from plants to all the animals in the ecosystem through **food webs** and **food chains**. This food energy is needed for all living processes such as growth, movement, respiration and reproduction.

The arrows show how energy is moved through the food web. Because every plant and animal uses up energy to live, there is less and less energy to be passed on. This is why a food chain rarely has more than five links.

Decomposers also obtain their energy from the food web, when plants and animals die. These organisms help in the process of decay which is important to the cycling of nutrients.

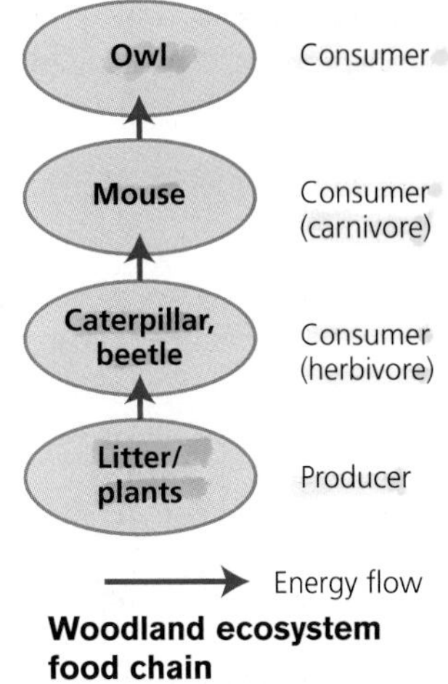

Woodland ecosystem food chain

Woodland ecosystem food web

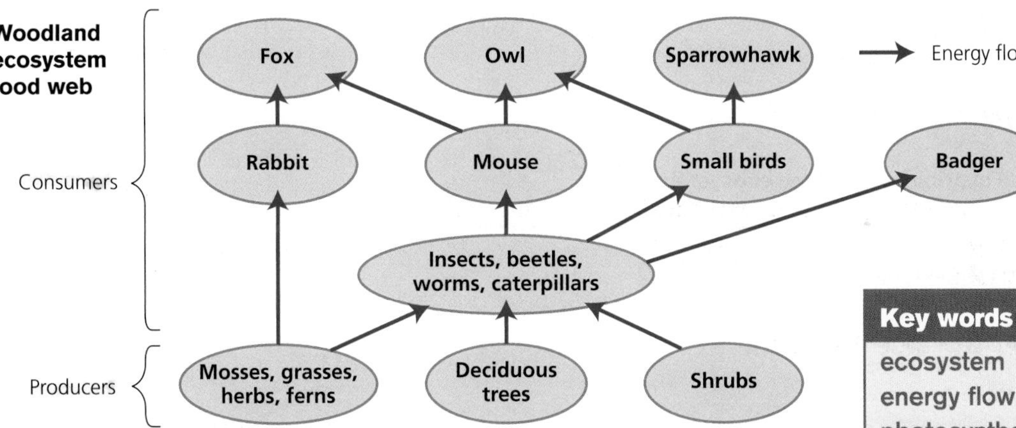

The impact of changing one component of the ecosystem

Any change to the balance of the ecosystem can trigger a reaction throughout the whole system. The reasons for change may be:

- **Natural:** for example, climate change with higher or lower temperatures or rainfall and therefore water availability or plant or animal diseases.
- **Human:** people have attempted to use or control the ecosystem. For example, cutting down trees and destroying habitats or replanting a woodland with a different type of tree.

Key words

ecosystem
energy flow
photosynthesis
primary producers
nutrient pools
biomass
litter
soil
leaching
food web
food chain
decomposers

If a producer, such as a tree, is destroyed, then there may be no food source for a particular type of insect and therefore no food source for the mice and small birds that feed on the insect and in turn for the owl, fox and sparrowhawk at the top of the food chain.

Temperate deciduous woodland

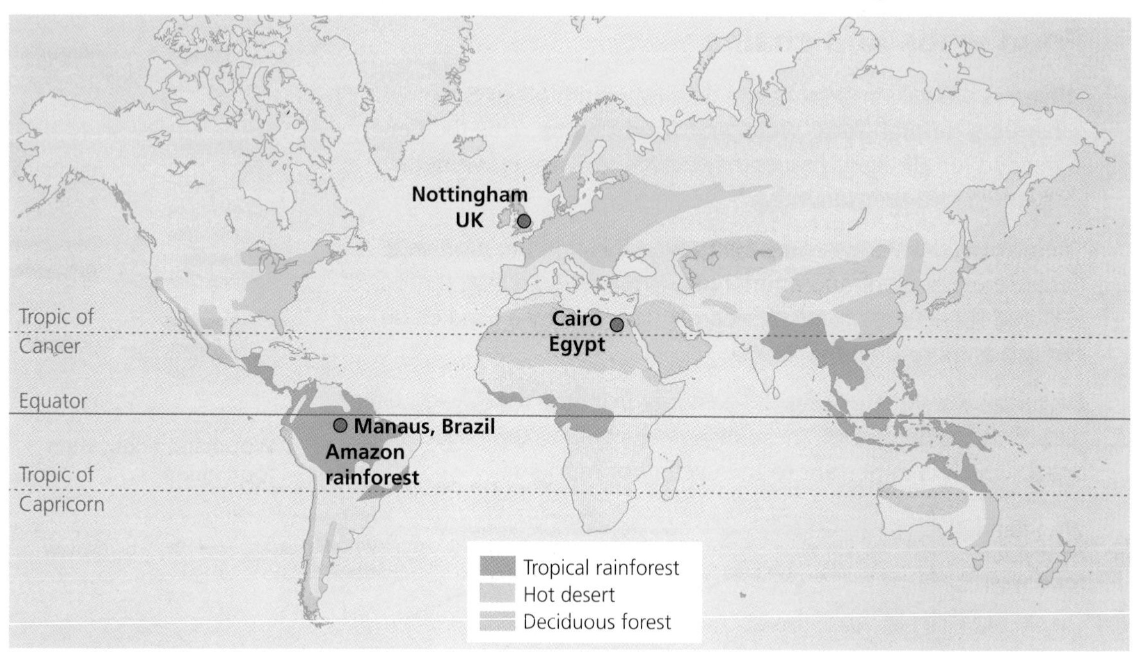

Global distribution of three ecosystems

Characteristics of climate

These areas have marked seasons with cold winters and warm summers

Precipitation is well distributed throughout the year, with no obvious wet season

There is a relatively long growing season of up to seven months

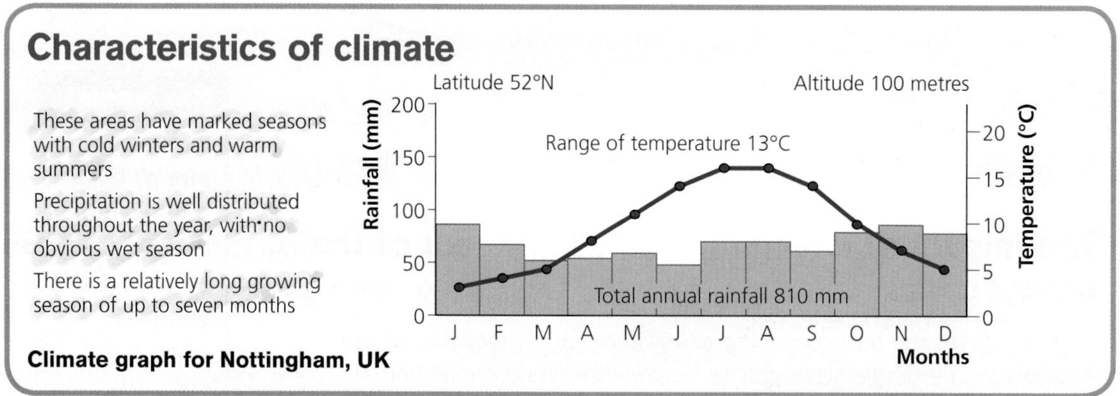

Climate graph for Nottingham, UK

Adaptation to the climate and soils

The trees and shrubs are adapted to the climate and shed their leaves in winter. In other words they are **deciduous**. This helps to reduce water loss and is a reaction to the colder temperatures and reduction in light levels during winter.

The leaves are generally wide and flat and do not need protection against water loss as rain falls throughout the year.

Characteristics of soils

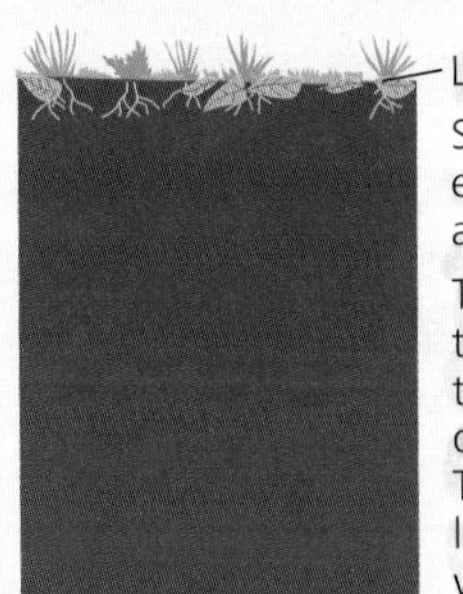

Leaf litter

Soils are generally 'brown earth' type, 30 cm deep and a reddish-brown colour

They are darker brown near the surface where most of the humus from the plentiful decaying matter is found. The trees are deciduous and lose their leaves in the winter providing a deep litter

There is no sharp division between the surface layer and the lower layers, which gradually become lighter brown

There is some leaching of minerals but this is slow

These are generally deep, rich soils well suited to agriculture

Brown earth

Characteristics of vegetation

The natural vegetation is broad-leafed deciduous trees such as oak, ash, beech and elm in Europe and maple in North America.

The forest is made up of four layers. There is a **stratification** of the vegetation:

The tree or **canopy** layer — beech, oak and ash up to 40 m high

The shrub layer or **sub-canopy** saplings and shrubs such as hazel and holly

The herb layer — non-woody plants such as brambles, ferns, grasses, bluebells and ivy

The ground layer close to the surface — fungi, mosses and lichens

Temperate deciduous woodland

Case study: *the use of a temperate deciduous woodland*

Padley Gorge Woods, Derbyshire

Factfile

- Native species were oak, hazel and silver birch. The Duke of Rutland planted Scots pine, sycamore and beech to give variety and colour in winter.
- Designated a Site of Special Scientific Interest due to the variety of species and age.
- Owned by the National Trust and lie in the heart of the Peak District National Park.

Location of Padley Gorge Woods

Use of the woodland in the nineteenth century

- As part of the Duke of Rutland's shooting estate.
- Charcoal burning using **coppiced wood**. Charcoal was made by burning the coppiced wood in a burner for use in the steel industry.
- Bark from trees was used to tan leather in Manchester.
- Sheep grazing up until the 1980s, which prevented the natural re-growth of many species.

Use of the woodland today

- Recreation, scientific interest and research.

How this temperate deciduous woodland is managed

- Cutting down the non-native species and fencing to stop sheep eating the young trees.
- This allows more light and encourages young trees, shrubs and flowering plants to grow.
- Damage by visitors has been reduced and the woods have more native plants and animals.

Key words

deciduous
stratification
canopy
sub-canopy
coppiced wood

Tropical rainforest

There are more plant and animal species, including mammals, birds and insects, in the rainforest than anywhere else on earth. The forest is very dense and plant growth is vigorous. The few indigenous peoples who live in the forest live sustainably in harmony with their environment.

Characteristics of climate

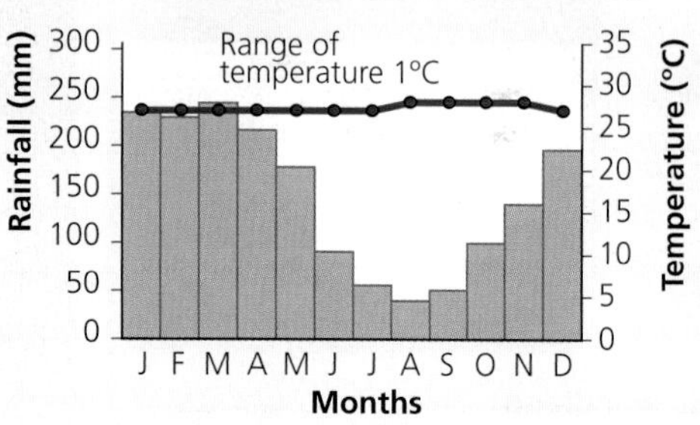

Temperatures are hot throughout the year — about 26 or 27°C.

Rainfall is heavy and mainly convectional in origin. Total rainfall is 1,773 mm.

It rains most days, usually in the afternoon.

No months are dry but there is a drier period between June and September.

There are no seasons.

It is always hot and wet.

Climate graph for Manaus, Brazil

Characteristics of soils

Despite the lush vegetation the rainforest **latosol** soils are surprisingly infertile

Dead leaves decompose rapidly on the surface in the hot and humid conditions

The soils are very deep but only a thin surface layer contains organic material and nutrients from the decomposed material. This is the only fertile layer

The heavy rainfall is much higher than the evapotranspiration so the rainwater flows downwards and out of the soil, taking with it the minerals and nutrients — this process is known as **leaching**

Below the fertile layer is a deep red and yellow infertile soil that gets its colour from the oxides of iron and aluminium which remain after the other minerals have been leached out

Rainforest latosol soil

Characteristics of the vegetation

Emergents or forest giants, 50 m or taller, e.g. kapok. These trees are usually supported by buttress roots

The **canopy** is a dense layer forming almost complete cover. Trees 20–30 m tall include many hardwoods such as mahogany

The **understorey** is between the canopy and the forest floor. This dark, dank area contains saplings between the trunks of larger trees

Large numbers of creepers and lianas entwine themselves around the tree trunks

The **forest floor** is covered in a deep litter of fallen leaves and branches

Structure of the tropical rainforest

Adaptation to the climate and soils

The leaves of the tallest trees grow at the top of the tree in the sunlight and have a leathery surface to withstand the powerful rays of the sun

The tallest trees, the emergents, grow very quickly to out-compete other trees to reach the sunlight

The leaves have 'drip tips' which are shaped to shed the water during heavy rains

Epiphytes are plants that grow in forks and on branches high in the canopy to reach the sunlight, there is no soil here so they live on nutrients from the rainfall and air

The leaves have a flexible base so they can turn to face the sun

Lianas and other climbing plants clamber up the trees to reach the light

In the lower layers, ferns and other plants make maximum use of any light that reaches them and can tolerate dark places

The tallest trees have buttress roots to support their tall trunks and shallow roots

Plants have shallow spreading roots as the fertile soil layer is so thin

Shrubs quickly take advantage of any gap in the canopy but generally there are few growing on the dark forest floor

Adaptation within a rainforest

Case study: a tropical rainforest

The exploitation of the Amazon Basin

20% of the Amazon rainforest has been cut down in the past 40 years.

Causes of deforestation

- Commercial logging of hardwood trees such as mahogany for cellulose, plywood, veneers and planks.
- Roads built by loggers. Settlers move in and burn areas of forest to grow crops.
- Cattle ranching: huge areas have been cleared for grazing cattle.
- Commercial agriculture: soybeans, cotton and corn.
- Mining, hydroelectric schemes and oil exploration.

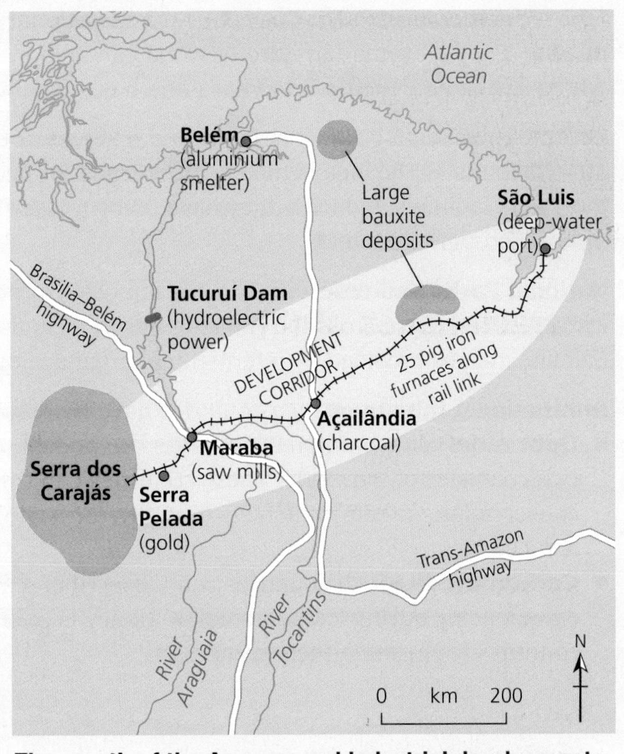

The mouth of the Amazon and industrial developments

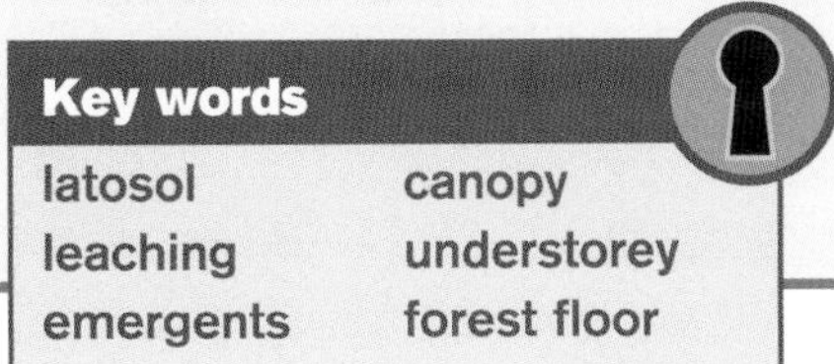

Effects of rainforest deforestation

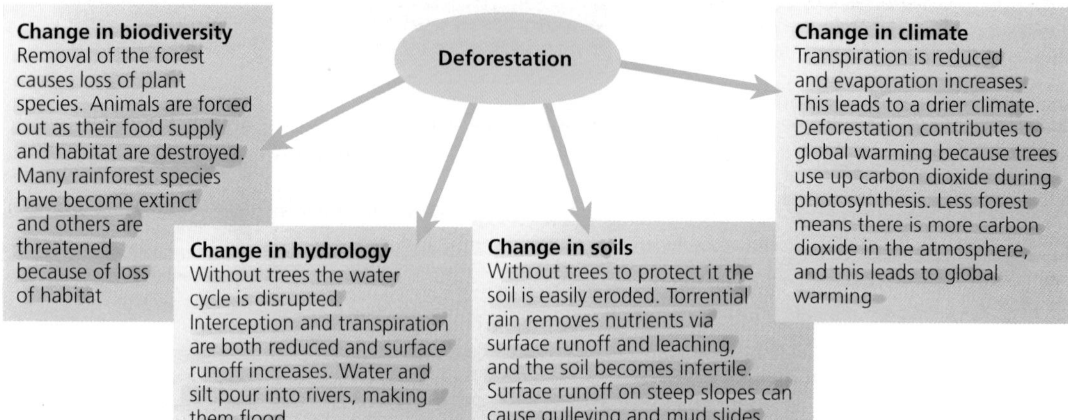

Deforestation
Change in biodiversity
Removal of the forest causes loss of plant species. Animals are forced out as their food supply and habitat are destroyed. Many rainforest species have become extinct and others are threatened because of loss of habitat
Change in hydrology
Without trees the water cycle is disrupted. Interception and transpiration are both reduced and surface runoff increases. Water and silt pour into rivers, making them flood
Change in soils
Without trees to protect it the soil is easily eroded. Torrential rain removes nutrients via surface runoff and leaching, and the soil becomes infertile. Surface runoff on steep slopes can cause gulleying and mud slides
Change in climate
Transpiration is reduced and evaporation increases. This leads to a drier climate. Deforestation contributes to global warming because trees use up carbon dioxide during photosynthesis. Less forest means there is more carbon dioxide in the atmosphere, and this leads to global warming

Sustainable tropical rainforest development in the Amazon

Sustainable logging Small-scale schemes to produce timber in a way that does not destroy the forest. **Selective logging** involves identifying trees to be cut and taking only the marked trees out of the forest. This causes minimum damage to the surrounding forest.

FSC (Forest Stewardship Council) An international organisation that promotes sustainable forestry. People buying furniture or wood can help by making sure that it is from a forest where sustainable methods are used and has a label with the FSC logo.

Ecotourism Creating National Parks and reserves not only protects the rainforest but also attracts tourists. The idea behind ecotourism is for people to visit the rainforest but not disturb the plants, animals and birds they have come to see. Visitors stay in lodges and travel on foot or boat through the forest.

National Parks and reserves There is a mosaic of protected parks, reserves, and conservation areas across the Amazon Basin, including the Jau Rainforest Park in Brazil. Satellite images provided by international organisations pinpoint illegal incursions.

International initiatives Rich countries have two major initiatives:

- **Debt relief** Many rainforest countries are in debt to the rich countries of the world. These countries can trade the conservation of their rainforests in return for having their debt reduced.
- **Carbon credits** Rich countries can 'offset' their carbon emissions by buying 'carbon credits'. Money is paid to poor countries to maintain their rainforest.

Key words

- sustainable logging
- selective logging
- ecotourism
- debt relief
- carbon credits

Hot deserts

Hot deserts are located in tropical and subtropical areas on the western side of land masses. The Sahara Desert is by far the largest, and bigger than all the others put together.

Animals survive by keeping cool and preventing water loss. Many rodents are small enough to hide under stones or in burrows during the day. Insects and reptiles are common in the desert, they have waterproof skins and produce little urine.

Characteristics of climate and soils

Climate

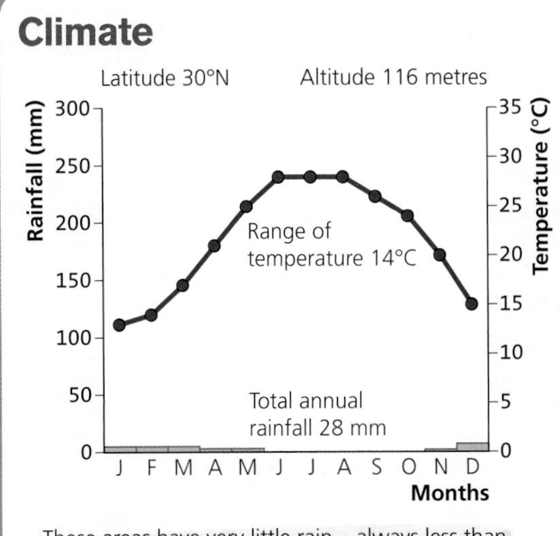

These areas have very little rain – always less than 250 mm per year. In some places, it may not rain for months or even years

As well as being dry it is also very hot. Average summer temperatures may reach 30°C but some days it may be as hot as 50°C in the shade

Temperatures fall quickly at night because there are no clouds. This leads to a large difference between day and night time temperatures — a large **diurnal temperature range**

Climate graph for Cairo, Egypt

Soils

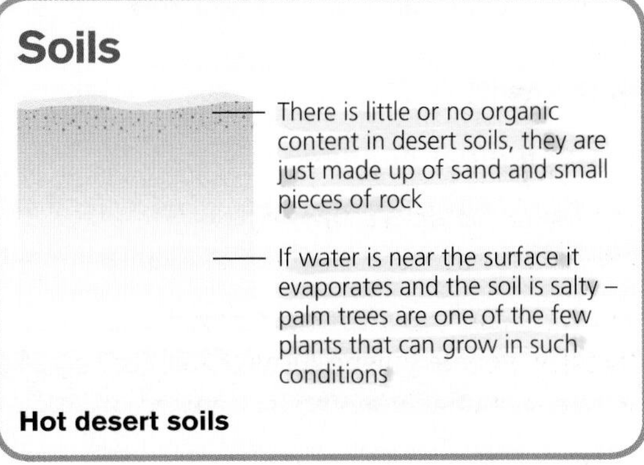

There is little or no organic content in desert soils, they are just made up of sand and small pieces of rock
If water is near the surface it evaporates and the soil is salty – palm trees are one of the few plants that can grow in such conditions

Hot desert soils

Key words

diurnal temperature range
succulent
nomadic pastoralists
desertification
bunds

Adaptation to the climate and soils

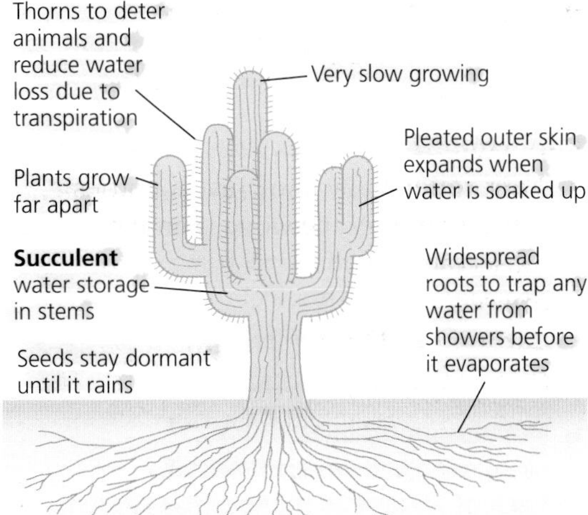

How the giant saguaro cactus has adapted to desert conditions

Case study: a hot desert in a rich part of the world

The Western Desert, USA

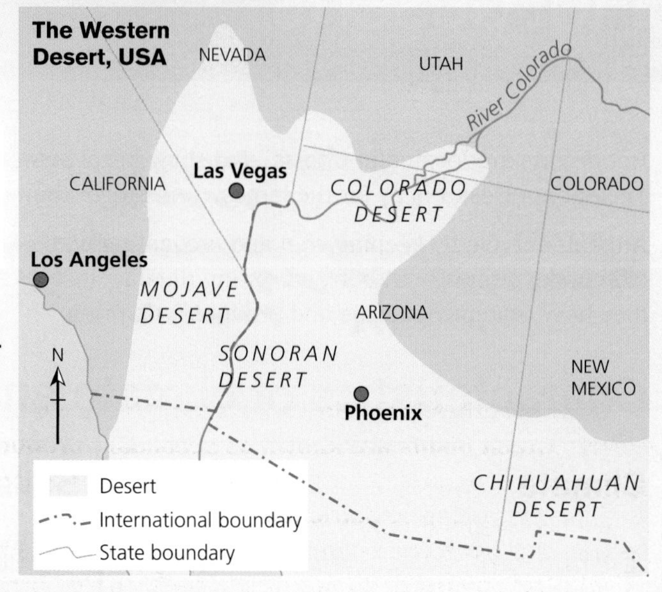

Factfile

- Population is low.
- Many tourists visit.
- There are three deserts in this area.

Opportunities

Tourism Dry sunny weather and scenery.

Commercial farming Using irrigation from underground aquifers.

Mining Copper, uranium, lead, zinc and coal.

Retirement

Management challenges

Water The aquifers are being emptied faster than they are replenished.

Tourism Is damaging the desert ecosystem. Planning is required for sustainable future settlement and protecting wildlife.

Case study: a hot desert in a poorer part of the world

The Thar Desert, India

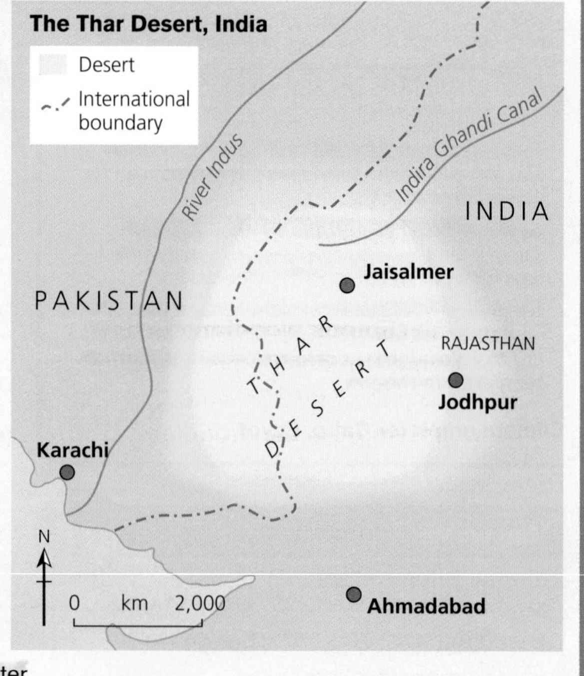

Factfile

- Temperatures can reach over 50°C and rainfall is less than 250 mm per year.
- Population densities are high for a desert – over 71 people per square kilometre.

Opportunities

- **Agriculture** sorghum, pulses and oilseeds. When fodder stores run out, **nomadic pastoralists** move sheep, camels and goats to find fresh grazing.
- **Mining** limestone and gypsum.

Management challenges

- **Population pressure** leading to overgrazing and **desertification**.
- **Lack of water and money.** Plans include planting new varieties of fast-growing trees, building small dams or **bunds** to conserve water.

Test yourself

1 Make a copy of the diagram and label it carefully.

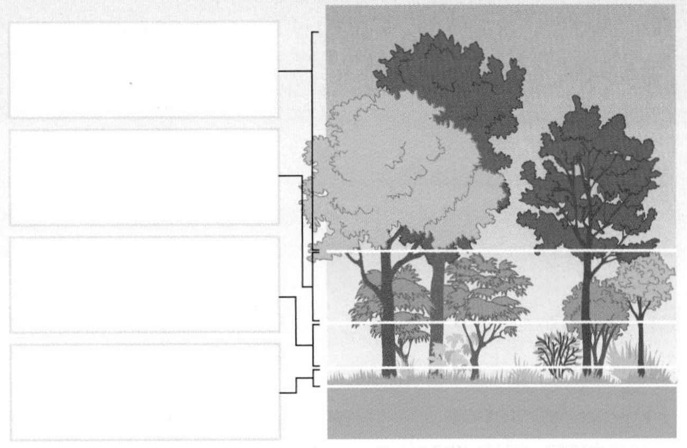

2 State whether the following statements are true or false.
   (a) Green plants are known as secondary producers.
   (b) All ecosystems have two processes – the flow of energy through the system and the cycling of nutrients.
   (c) Nutrients can be lost to the system due to leaching.
   (d) Decomposers do not obtain their energy from the food web.
   (e) Tropical rainforest soils are very fertile.
   (f) Deforestation can disrupt the water cycle.
   (g) Humus is inorganic.
   (h) Emergents grow between the canopy and the forest floor in a tropical rainforest.

3 List the ways in which the tropical rainforest ecosystem is being destroyed.

Exam tip

The final question often asks for more than one aspect, for example the last part of the higher tier question asks for use *and* management. Check you have answered every aspect.

Examination question

Foundation tier:

(a) For a deciduous woodland ecosystem you have studied:
   (i) Name the deciduous woodland ecosystem you have chosen.
   (ii) Describe how the woodland is used. *(3 marks)*
   (iii) Explain how the woodland is managed. *(4 marks)*

Higher tier:

(b) For a deciduous woodland ecosystem you have studied:
   (i) Name the deciduous woodland you have chosen.
   (ii) Describe how the woodland is used and how it is managed for these uses and conservation. *(7 marks)*

River processes

Rivers are important agents of **erosion**, **transport** and **deposition**. The shape of river valleys changes as rivers flow downstream due to the varying importance of each river process.

Erosion

Rivers erode in four main ways:

1. **Hydraulic action** The power of running water undercuts the banks and erodes the bed.
2. **Abrasion** or **corrasion** Rocks and pebbles carried by the river erode material from the bed and sides.
3. **Attrition** Rocks and pebbles carried by the river bang into each other and break up into smaller pieces.
4. **Solution or corrosion** Some minerals from rocks dissolve in the river.

Transport

Rivers transport an enormous load. This material can be carried in four ways:

1. **Traction** Boulders and rocks are dragged or rolled along the river bed.
2. **Saltation** Smaller-sized particles are bounced along the bed.
3. **Suspension** Fine particles of silt or clay are held in the water.
4. **Solution** Dissolved minerals.

Suspended load

Bed load Pebbles, rocks and boulders are rolled or bounced along the channel bed

Dissolved load Material dissolved in the water which cannot be seen

Stones become smaller and rounded as they are moved due to attrition

Cross-section through a river

Deposition

A river drops its load when it no longer has enough energy to carry it. Larger, heavier material is deposited first, usually higher up the river. Pebbles, gravel, sand and silt are deposited in the middle and lower course. The dissolved load is not deposited but is carried out to sea.

The long profile and changing cross profile

The **long profile** shows how the river changes in height along its course. The **cross profile** also changes; the valley sides become less steep and the river channel becomes deeper and wider.

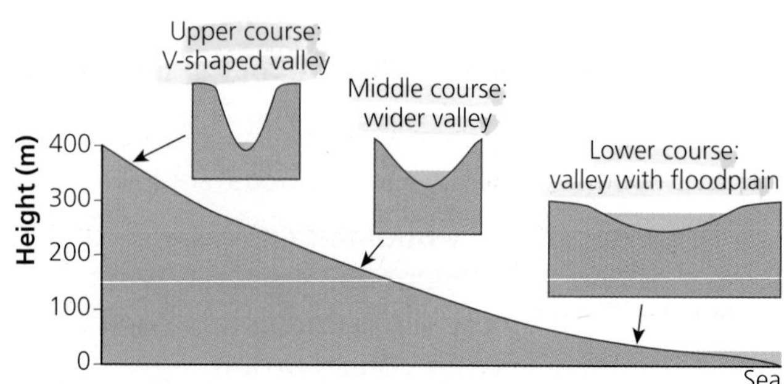

Long profile of a river showing changing cross-sections

River landforms

Landforms resulting from erosion

Waterfalls and gorges are found in the upper course of the river where downward (vertical) erosion predominates.

As the waterfall erodes, it moves slowly upstream, leaving a steep gorge on the lower side of the falls.

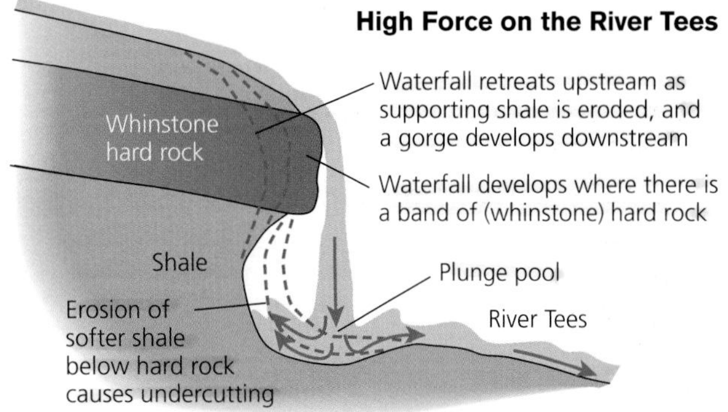

High Force on the River Tees
Whinstone hard rock
Shale
Erosion of softer shale below hard rock causes undercutting
Waterfall retreats upstream as supporting shale is eroded, and a gorge develops downstream
Waterfall develops where there is a band of (whinstone) hard rock
Plunge pool
River Tees

Landforms resulting from erosion and deposition

Meanders and oxbow lakes are found in the middle and lower course of the river. Lateral erosion is more important than downward erosion and some deposition also takes place.

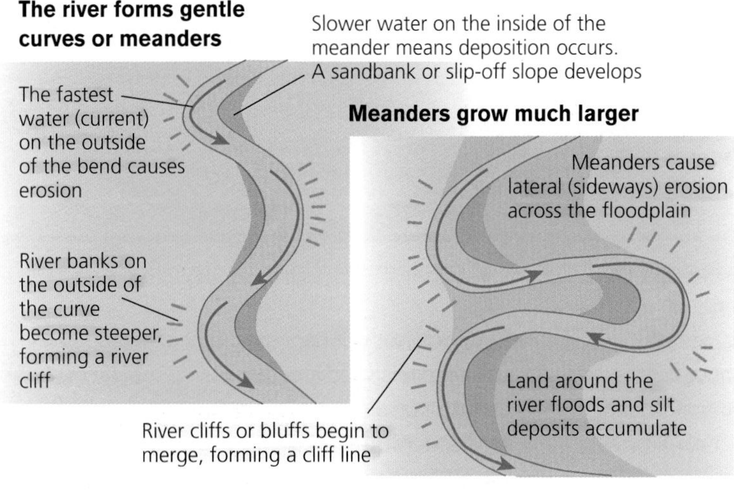

The river forms gentle curves or meanders
The fastest water (current) on the outside of the bend causes erosion
River banks on the outside of the curve become steeper, forming a river cliff
Slower water on the inside of the meander means deposition occurs. A sandbank or slip-off slope develops
Meanders grow much larger
Meanders cause lateral (sideways) erosion across the floodplain
Land around the river floods and silt deposits accumulate
River cliffs or bluffs begin to merge, forming a cliff line

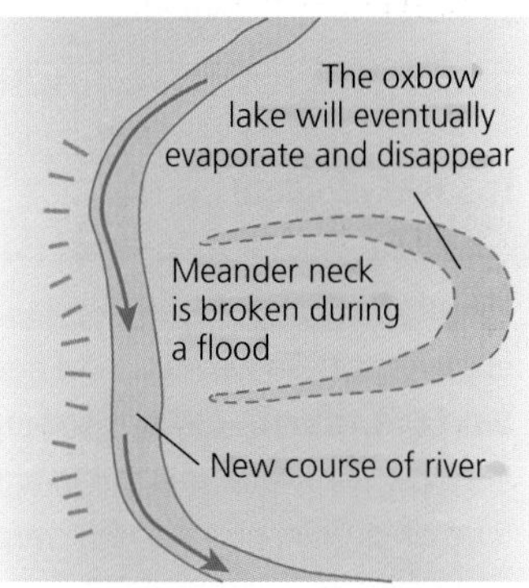

Meander is cut off from the main channel, usually during a flood. An oxbow lake is formed
The oxbow lake will eventually evaporate and disappear
Meander neck is broken during a flood
New course of river

Landforms resulting from deposition

The land surrounding the river in its lower course is subject to frequent flooding. When the river floods, sediment is deposited on the land around it with the smallest material being carried to the edge of the floodplain closest to the bluff line, and larger heavier material is deposited nearer the river channel. Mounds of sediment running alongside the river may develop, which are higher than the surrounding floodplain. These mounds or embankments are called leveés.

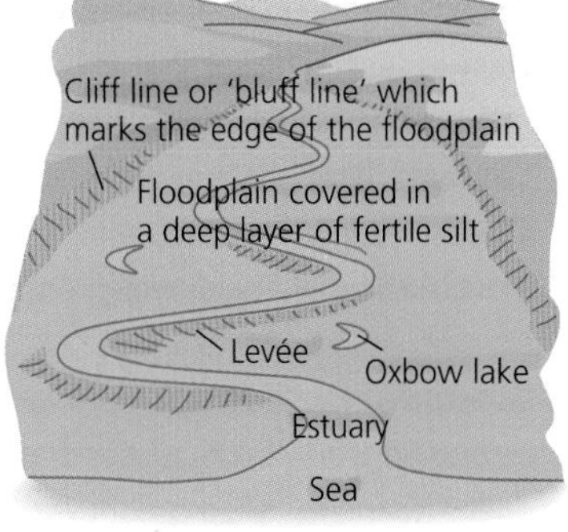

Cliff line or 'bluff line' which marks the edge of the floodplain
Floodplain covered in a deep layer of fertile silt
Levée
Oxbow lake
Estuary
Sea

River floodplain

The hydrological cycle

The hydrological cycle or the water cycle is the continuous movement of water between the land, the sea and the atmosphere.

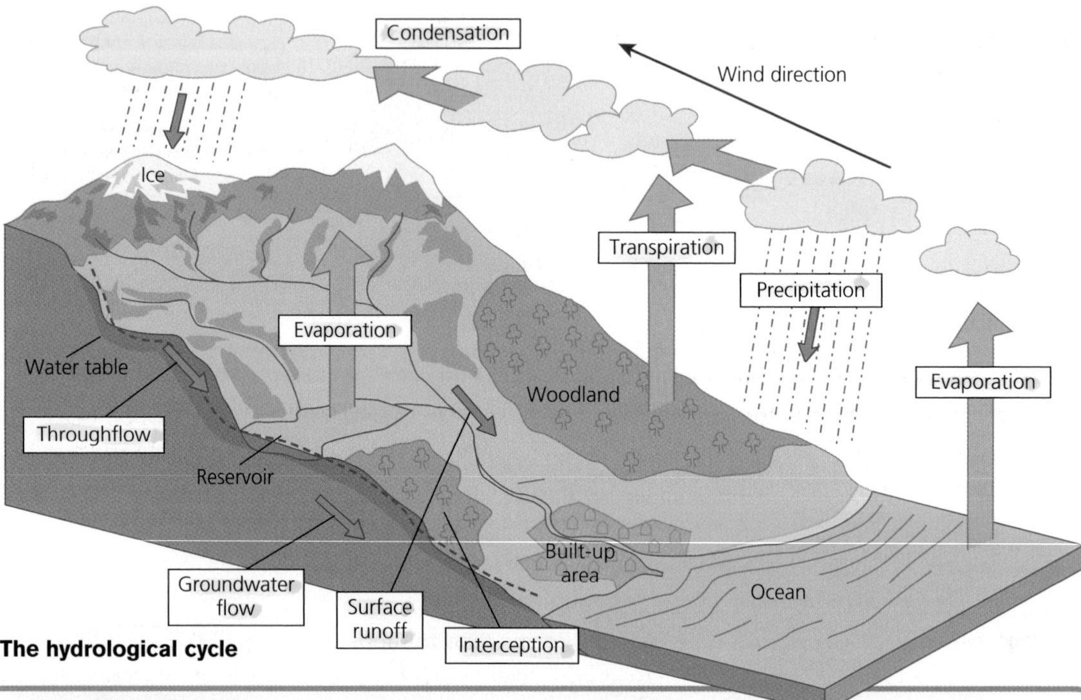

The hydrological cycle

- **Precipitation** – All water released from clouds such as rain, snow, hail, sleet and fog.
- **Surface runoff** – Water flowing across the surface. The water may be in a channel, such as a river or stream, or it may be overland flow when it makes its way across a field or down a roadway.
- **Interception** – When water collects on objects such as leaves or flat roofs.
- **Infiltration** – When water soaks into soil.
- **Throughflow** – When water soaks into soil and seeps through it towards a river or the sea.
- **Percolation** – The downward movement of water through soil into rocks.
- **Groundwater flow** – The movement of water below the water table. Water that is stored underground is called groundwater.
- **Evaporation** – When water that is heated by the sun becomes vapour and rises into the atmosphere. This may take place over land or sea.
- **Transpiration** – All plants lose water through their leaves. Transpiration is when this water returns to the atmosphere where it evaporates.
- **Evapotranspiration** – The term for the processes of both evaporation and transpiration.
- **Condensation** – When water vapour is cooled and turns into water droplets to form clouds.
- **Water table** – The upper level of saturated ground. It is not a 'table'. It is not even flat. The level is closer to the surface in winter when there is plenty of rain.

The storm hydrograph

The storm **hydrograph** shows river discharge, which is the amount of water in a river. You can see how **discharge** changes after a rainstorm.

Storm hydrographs show how a river reacts to heavy rainfall so they can be used to help planners to decide about flood prevention strategies.

A hydrograph is divided into two parts: **base flow**, which is water entering the stream from ground storage and **storm flow**, which is the water from the recent rainstorm. Base flow remains constant but the storm flow changes.

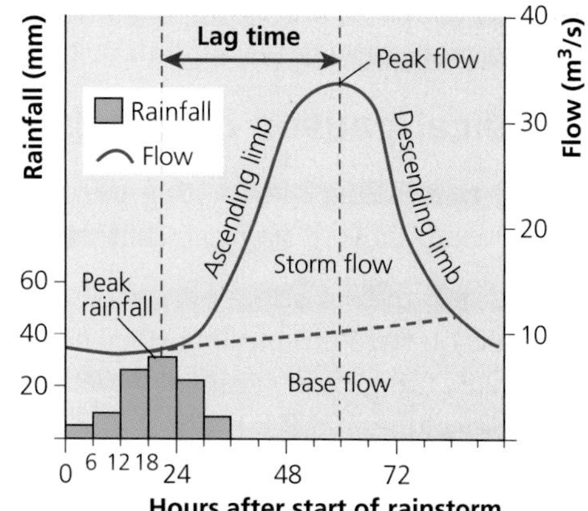

A storm hydrograph

The **ascending (or rising) limb** shows how quickly river discharge increases and then reaches **peak discharge**, which is the highest amount of water in the river. The time between peak rainfall and peak discharge is known as the **lag time**. The **descending (or falling) limb** shows how quickly river levels return to normal.

Factors affecting discharge

River discharge is affected by a number of factors:

- **Amount and type of rainfall** Heavy rainstorms lead to more surface runoff, increasing river discharge and flood risk, whereas light rain or drizzle is more likely to infiltrate.
- **Previous weather conditions** If there has been a lot of rain in recent weeks the ground may be saturated which will lead to increased surface runoff.
- **Relief** Steep slopes encourage surface runoff, partly because of the gradient but also because steep slopes tend to have less soil and so rainwater is less likely to infiltrate.
- **Rock type** Permeable or porous bed rock such as limestone or chalk allows more infiltration than impermeable rock such as sandstone and granite. If more water can infiltrate and be held as groundwater there will be less surface runoff and less risk of floods.
- **Land use** Land being used for growing crops or woodland is less likely to flood; trees and leaves intercept rain so it is less likely to reach the river, the ground surface is soft allowing infiltration; the water table will be lower because plants take up water from the ground.
- **Built-up areas** They have higher flood risk because the ground surface is covered by tarmac or concrete and is impermeable, which means there is little infiltration. Drains carry rainwater to the river very quickly leading to a rapid increase in discharge. Debris in the river may also increase the risk of rivers flooding.

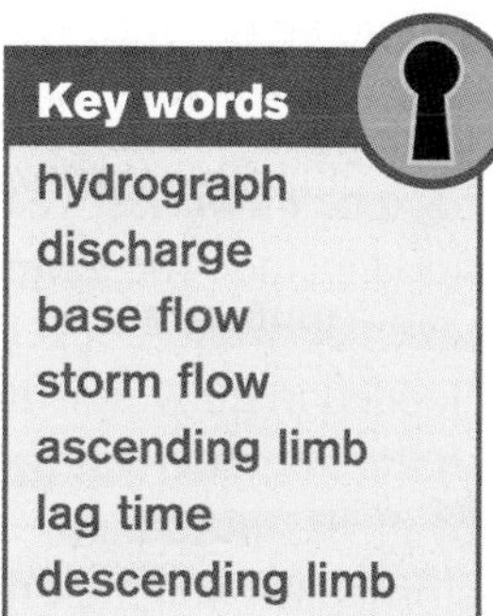

Key words
hydrograph
discharge
base flow
storm flow
ascending limb
lag time
descending limb

Flooding

If a river overflows its banks and inundates the surrounding land it is called a flood. Severe floods can destroy homes and infrastructure and cause enormous disruption to people's lives.

Physical causes of floods

Heavy rain falling over a long period of time can cause saturation of the ground and a rise in the water table. If no more rain can infiltrate, rivers fill and overflow their banks.

Heavy rain over a short period of time can cause floods particularly if the rain is intense and falls on hard ground that does not allow water to infiltrate. This can happen if hot weather has baked the ground. Rainfall flows quickly into rivers, which rise rapidly and overflow their banks resulting in a **flash flood**.

Rapid snow melt in late winter and spring may cause flooding as the amount of water flowing into the river is too much for the river to hold.

Relief Water runs down steep slopes quickly and this can contribute to flooding.

People's activities that can cause flooding

Urbanisation In recent years there has been more building on land beside rivers. This leads to the river's floodplain being covered in tarmac or concrete which are impermeable and increase surface runoff so more water flows into the river.

Deforestation If trees are removed there is little to break the intensity of rainfall hitting the ground, which may cause soil erosion. Eroded soil is washed into river channels and may be deposited in lowland areas reducing the capacity of the river and making flooding more likely.

River management Although the main aim of river management is to reduce the likelihood of flooding, in some situations it can lead to increased risk of flooding. For example, straightening a river channel and lining it with concrete can mean that an area further downstream becomes at greater risk of flooding because water reaches it more rapidly.

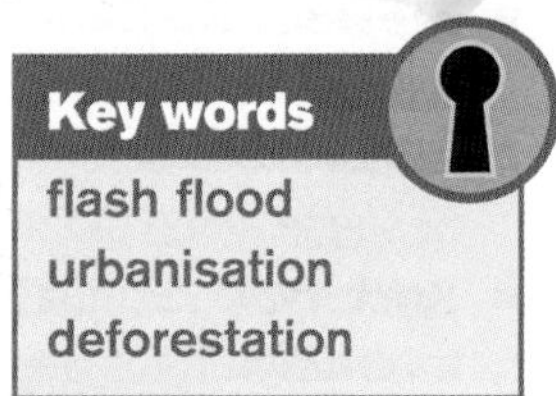

Flooding in the UK

During the last 20 years flooding in this country has become an almost yearly event. Rivers that frequently flood include the River Severn, the River Ouse and the River Derwent. Cockermouth in Cumbria was badly flooded in 2009. It is likely that flooding will increase in the future as a result of climate change caused by global warming because an increased number of storms will heighten flood risk.

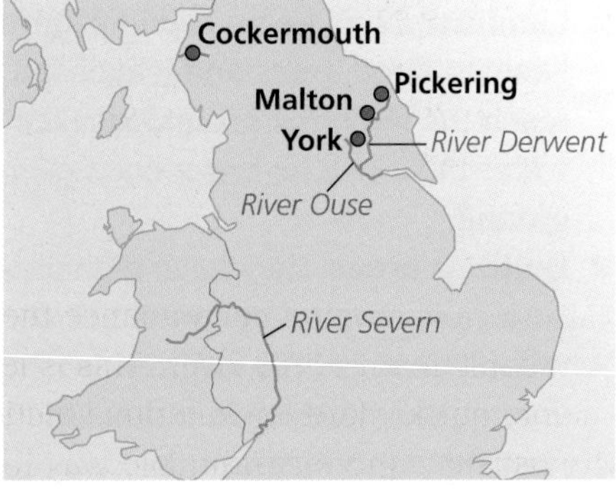

Flood control

Hard engineering is expensive to install and to maintain. It can also spoil the environment and impact on wildlife.

Integrated river management emphasises **sustainable** solutions to the problems of flooding. It combines some hard engineering with **soft engineering**. It considers the impact on people and the environment as well as the cost of the scheme.

Hard engineering

- Building embankments, also called 'dykes' or 'levées' to keep flood waters in the river.
- **Channelisation** Straightening and deepening the river so water flows away fast. The river will revert to its natural course unless the new channel is reinforced.
- Constructing dams and reservoirs in the upper sections of the river to trap water that can be released slowly. Reservoirs can be used for recreation or for generating hydroelectricity but these schemes are very expensive and lead to loss of land and sometimes displace a lot of people.

Embankments or dykes, built from rocks and soil and grassed over
River straightened and dredged
Meander infilled to create straighter river

Hard engineering solutions

Flood relief channel
Willow and alder
Natural river cliff or bluff
Embankments set back from river channel

A natural approach to flood control

Natural flood control (soft engineering)

- Setting embankments back from the channel and avoiding building on floodplains to allow flooding.
- Planting water-loving plants, such as willow and alder. This helps to lower the water table.
- Digging a flood relief channel so the river can cope with increased discharge when necessary. In normal conditions, the meandering river course is maintained.
- Providing general maintenance for rivers to keep the water flowing – for example cleaning rubbish and debris from the channel.
- Dredging channels to enlarge them.
- Planting trees (**afforestation**) in the upper catchment area. This reduces the amount of water reaching the river but changes the appearance and ecosystem of the area.

Natural flood control can enhance the environment and provide access for people to enjoy the river. This is less expensive to implement and maintain than hard engineering solutions. Soft engineering approaches are usually a more sustainable way to protect areas from floods.

Key words

hard engineering
integrated river management
sustainable
soft engineering
channelisation
afforestation

Case study: *flooding in a rich part of the world*

River Severn floods, 2007

The River Severn floods regularly for a number of reasons:

- The Cambrian Mountains where the river rises is an area of high rainfall.
- It has a large catchment area and several large tributaries channelling water into it.
- It flows across impermeable rock for much of its length so water cannot infiltrate and runs rapidly into the river.
- In recent years towns by the river have grown and floodplains have been used for new buildings increasing surface runoff.
- Flood defences in some places (such as Shrewsbury) increase the flood risk in other areas further downstream.

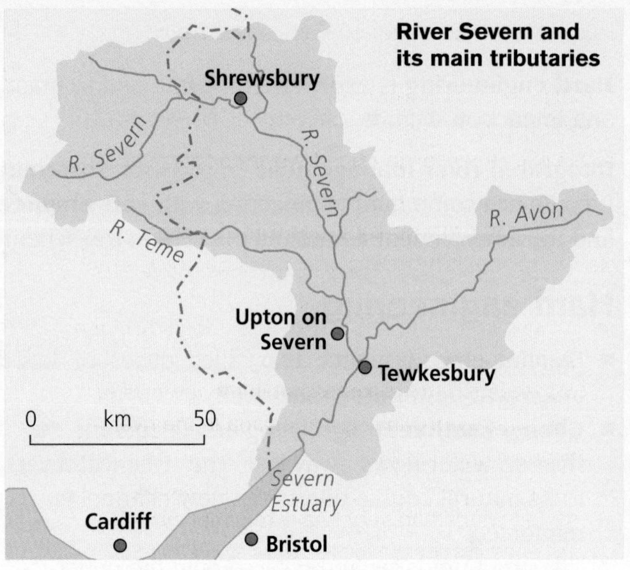

July 2007 floods

Causes

Floods along the River Severn were caused by heavy rain over a prolonged period resulting in the surrounding land becoming saturated and causing increased surface runoff. The river channel could not contain all the water and it overflowed its banks. Flooding was probably made worse by building on the floodplain, which would otherwise provide a natural overspill area for water from the river.

Effects

The floods caused enormous damage to properties along the river particularly in Upton on Severn and Tewkesbury, which was completely cut off.

- Three people died and thousands had to leave their homes.
- Many houses and businesses were badly damaged.
- Roads and railway lines were closed and stranded people had to be rescued by the emergency services.
- A water treatment centre at Tewkesbury and an electricity sub-station at Gloucester were closed leaving 350,000 people without water or electricity for several days.

Over a year later, some people were still unable to return to their homes. Farmers who lost their crops had no income and nothing to invest in their farms for the future, and some shops and businesses lost so much stock they had to close.

Response to flooding

The River Severn has extensive flood protection schemes in place, e.g. in Shrewsbury which was not affected by the 2007 floods.

Upton on Severn is protected by temporary flood barriers that are put in place when river levels start to rise. However, in July 2007 the barriers did not arrive in time as they were stuck on the motorway and this allowed flood waters to rise and inundate the town. A £3.6m scheme has been proposed which includes building a wall running 400 m along the river. This plan is controversial as it will take away the riverside views.

Tewkesbury has no engineered flood defences. Since 2007, local people have asked for proper flood defences but there are no plans for this. By 2026, just under 15,000 new homes are scheduled to be built in and around Tewkesbury. If these are built on the floodplain they are likely to increase the flood risk for the town.

Case study: *flooding in a poor part of the world*

Bangladesh

Bangladesh lies almost entirely on the Ganges Delta, and most of the country is flat and low lying. Almost all of it is lower than 12 m above sea level. Flooding occurs every year leaving behind valuable fertile silt. However, in recent years floods have become bigger and lasted longer.

Flooding is caused by a combination of:

- the **monsoon** climate which brings heavy rain between June and October
- snow melt from the Himalayas, particularly in the summer
- the effects of **deforestation** in the foothills of the Himalayas causing deposition in the river channel downstream
- tropical storms and cyclones, which cause strong winds and very heavy rain that can increase discharge in the rivers and cause floods

Effects of flooding in Bangladesh – 2004 and 2007

- Deaths, serious injury and loss of life.
- Loss of and damage to houses causing many people to become homeless.
- Major disruption to transport as roads and railways were flooded.
- Damage to community buildings such as schools and hospitals.
- Destruction of food crops, particularly rice, so food supplies disrupted for months.
- Rapid spread of **water-borne diseases** including dysentery and diarrhoea. More deaths usually caused by disease rather than from the flood itself.

Bangladesh is particularly vulnerable to **climate change**. It is low lying and at risk from rising sea levels – estimates suggest there could be 25 million climate refugees in the future. Climate change also increases the frequency and severity of tropical storms.

Response to flooding in Bangladesh

Bangladesh has to rely on aid from other countries and from non-government organisations (NGOs) such as Oxfam and Save the Children. Short-term emergency aid brings medical and food supplies to the flooded area and long-term aid is used to help the country rebuild its infrastructure.

In the past, earth embankments up to 7 m high have been constructed along 6,000 km of the River Ganges. However, these have not been successful because:

- the **embankments** are easily breached
- they tend to create a false sense of security for people living nearby
- they make it difficult for flood water to flow back into the river and prolong floods

Now the emphasis is on reducing damage rather than on controlling floods. This is being done by:

- having better **flood warning** systems
- building **flood shelters** on stilts to provide somewhere safe for people to go
- planting vegetation along the banks of the river to increase interception and reduce the amount of groundwater
- dredging river channels and opening up abandoned channels to speed the flow of water away from the area

In the longer term it is important for Bangladesh to work with India and Nepal to develop an integrated plan to tackle the problem of flooding in the Ganges–Brahmaputra Basin.

Key words

monsoon
deforestation
water-borne diseases
climate change
embankment
flood warning
flood shelter

Water supply and demand in the UK

We each use about 150 litres of water every day. Two-thirds of our water comes from **surface** sources (reservoirs and rivers) and the rest from groundwater. Sources vary by region, but in London and the South East, groundwater accounts for around 70% of the water supply.

There is enough water to supply the population of Britain. However, there is more water in areas of low population density, such as Scotland and northwest England, than in the south and east where most people live and where demand for water is increasing.

Demand for water is growing because:

- the population is increasing
- more people live alone so more houses are needed
- greater affluence means people use more water at home, e.g. in dishwashers
- industries use more water every year

Supplying enough water in the future could be a problem, especially if climate change brings longer, hotter and drier summers.

In the past, attempts to increase water supply focused on building large **reservoirs** such as Carsington Water in Derbyshire and Kielder Water in Northumberland.

The emphasis now is on meeting future demands in a **sustainable** way through small-scale, local schemes and not through large-scale **transfers**.

- Water companies must reduce the amount of water lost due to leaking pipes.
- Many houses now have their own meter reducing demand by about 10%; this is compulsory in areas of serious water stress.
- Modern equipment such as washing machines, dishwashers and toilets are designed to use less water.
- Education and advertising encourages people to use less water, e.g. by having a shower instead of a bath.

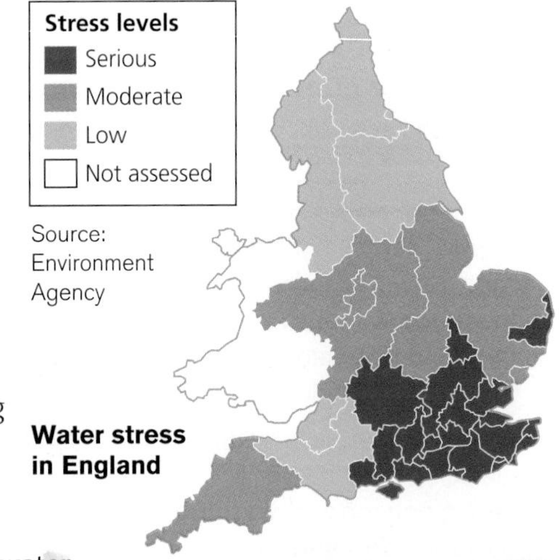

Water stress in England

The Environment Agency produced this map in 2007 using data of current and forecast demand, and estimated population growth. It identifies 21 areas where demand may be greater than supply (shown as 'serious' on the map) and these areas are said to have serious **water stress**. All are in south and east England, including London.

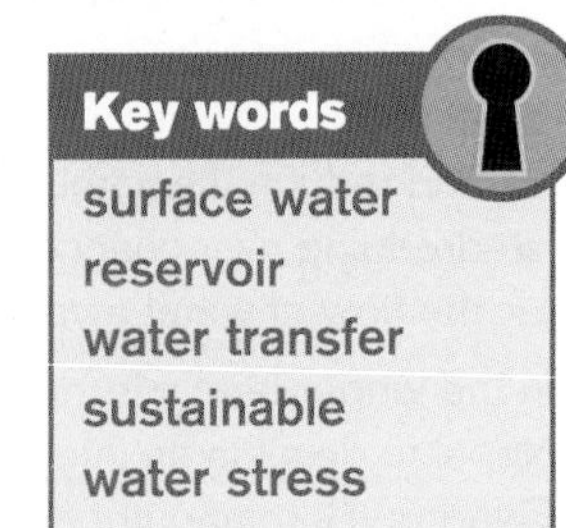

Key words
surface water
reservoir
water transfer
sustainable
water stress

Case study: *a reservoir*

Carsington Water, Derbyshire

Carsington Water was built by Severn Trent Water Authority in 1991 to improve water supply. The site at Carsington was chosen because:

- it is on impermeable rock
- there was enough water from local streams to fill the reservoir
- it didn't cause too much disruption to local people (only two farms were flooded)

A huge earth dam was built to hold back the water in the reservoir. All the clay and rock used for the dam wall was dug out of the valley, which later became the reservoir. The stone used for facing the outside of the wall was obtained from local quarries. The project cost £107 million.

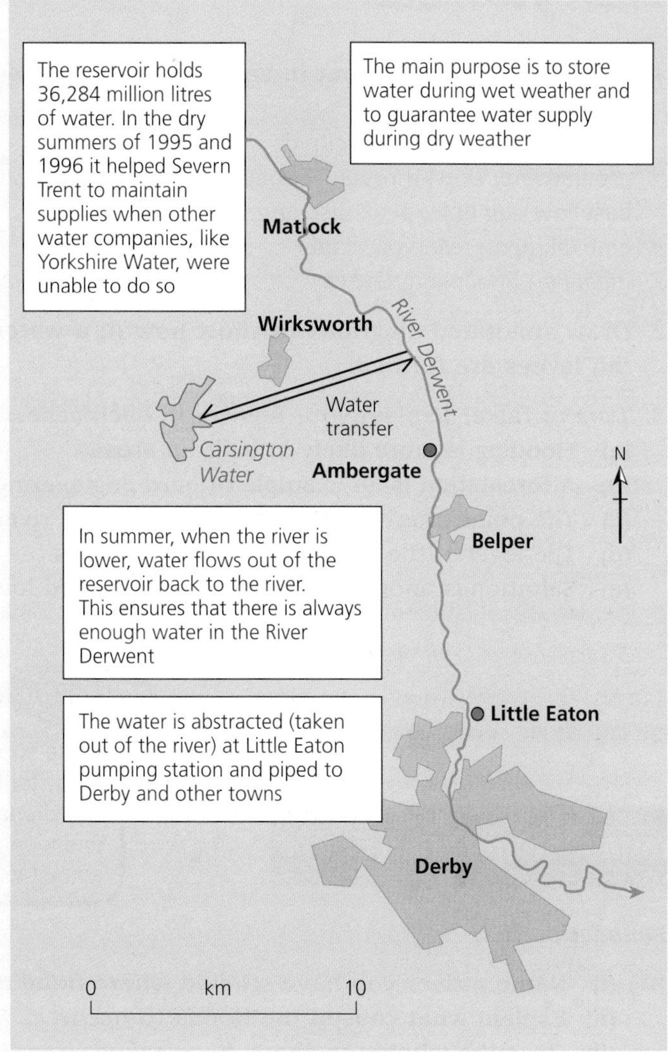

The reservoir holds 36,284 million litres of water. In the dry summers of 1995 and 1996 it helped Severn Trent to maintain supplies when other water companies, like Yorkshire Water, were unable to do so
The main purpose is to store water during wet weather and to guarantee water supply during dry weather
Matlock
Wirksworth
River Derwent
Water transfer
Carsington Water
Ambergate
N
Belper
In summer, when the river is lower, water flows out of the reservoir back to the river. This ensures that there is always enough water in the River Derwent
Little Eaton
The water is abstracted (taken out of the river) at Little Eaton pumping station and piped to Derby and other towns
Derby
0
km
10

The Carsington water management scheme

Socio-economic impacts

- The reservoir provides a reliable source of water for thousands of people.
- It has reduced the risk of flooding in the River Derwent valley.
- There are activities for visitors (over 1 million people a year), including walks, cycling, water sports, fishing, a restaurant and a playground.
- Jobs have been created, initially in building the reservoir and now looking after visitors.

Environmental impacts

- There is a completely new landscape – agricultural land has been replaced by the reservoir.
- The ecosystem has changed, but the water has attracted a wide variety of wildlife.
- A new road has been built for access. This means extra traffic and some pollution.
- Visitors can damage the environment, e.g. by dropping litter and causing footpath erosion.

Test yourself

1 **Which is the odd one out in each of these groups of words? Explain why in each case.**
precipitation, infiltration, clouds
groundwater flow, throughflow, surface runoff
baseflow, lag time, peak discharge
embankment, reservoir, canal
meander, floodplain, levées

2 **Draw annotated diagrams to show how (i) a waterfall (ii) an oxbow lake and (iii) levées are formed.**

3 **True or false? Explain your answer in each case.**
(a) Flooding is more likely in built-up areas.
(b) Afforestation is an example of hard engineering.
(c) The point where a tributary joins the main river is called a confluence.
(d) The level of the water table rarely changes.
(e) Saltation is another term for the suspended load in a river.

Exam tip

Make sure you have learnt a case study for the final part of exam questions. You need to know the name and location of your case studies and some facts about each one.

Examination question

Foundation tier:

(a) (i) Name a river you have studied where flooding has caused problems.
(ii) Explain what caused the floods to occur.
(iii) Describe what steps have been taken to manage the flooding. *(8 marks)*

Higher tier:

(b) Name a river you have studied where flooding has caused problems.
Describe how the river has been managed to reduce the flood risk.
To what extent are these protection measures sustainable? *(8 marks)*

Global distribution of ice

The last ice age

The last **ice age** is also called the **Pleistocene**. It lasted for about 2 million years and finished about 10,000 years ago. Even during an ice age temperature varies. The times when temperatures fall and ice advances are called **glacial periods** and periods that are warmer and ice retreats are called **interglacials**. Evidence from ice cores and deep-sea sediments suggest that there may have been as many as 20 cold periods or glacials during the Pleistocene period. Interglacial periods were at least as warm as today's climate.

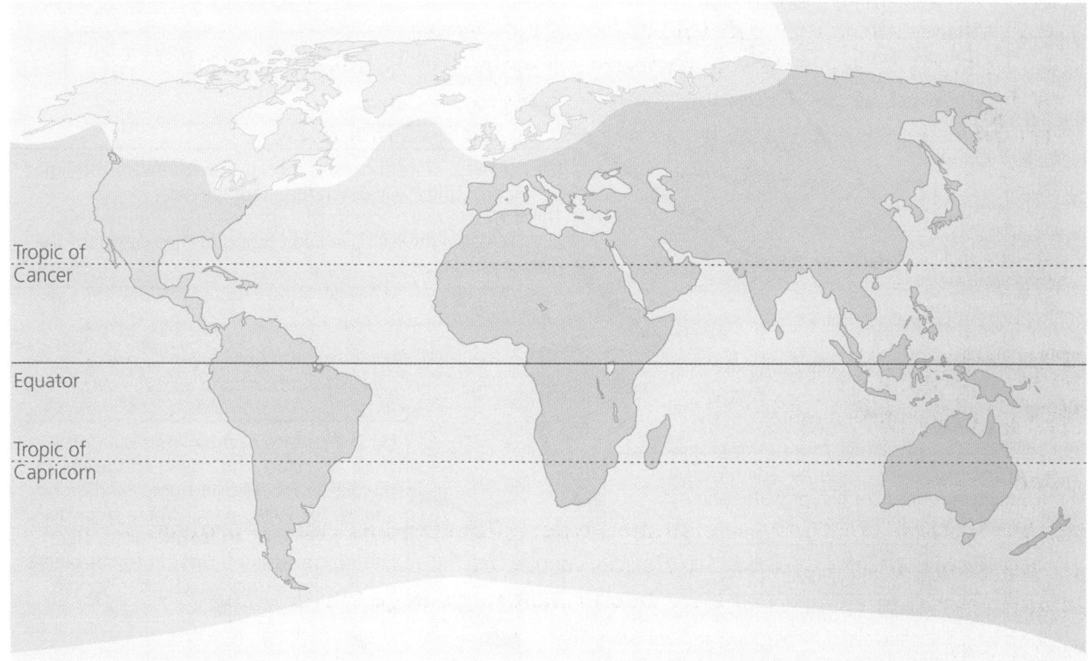

The map shows the extent of ice cover 18,000 years ago when large parts of North America and Europe, including most of the British Isles, were covered by ice sheets

Present day ice cover

There are two large **ice sheets** in the world covering Antarctica in the south and Greenland and Arctic Canada in the north. There is evidence that both these ice sheets are melting due to global warming.

Smaller areas of ice are called **ice caps** and these are found in highland regions such as the Alps and Himalayas. Individual **glaciers** spread out like fingers from ice caps and are still found in every continent in the world.

Key words

- ice age
- Pleistocene
- glacial period
- interglacial
- ice sheet
- ice cap
- glacier

A glacier

When snowflakes thaw and refreeze they form a whitish grain-like snow called **firn**. The weight of more snow falling on top forces out air, and the firn granules merge to form bluish **glacier ice**. This ice eventually moves down the mountainside and forms a **glacier**.

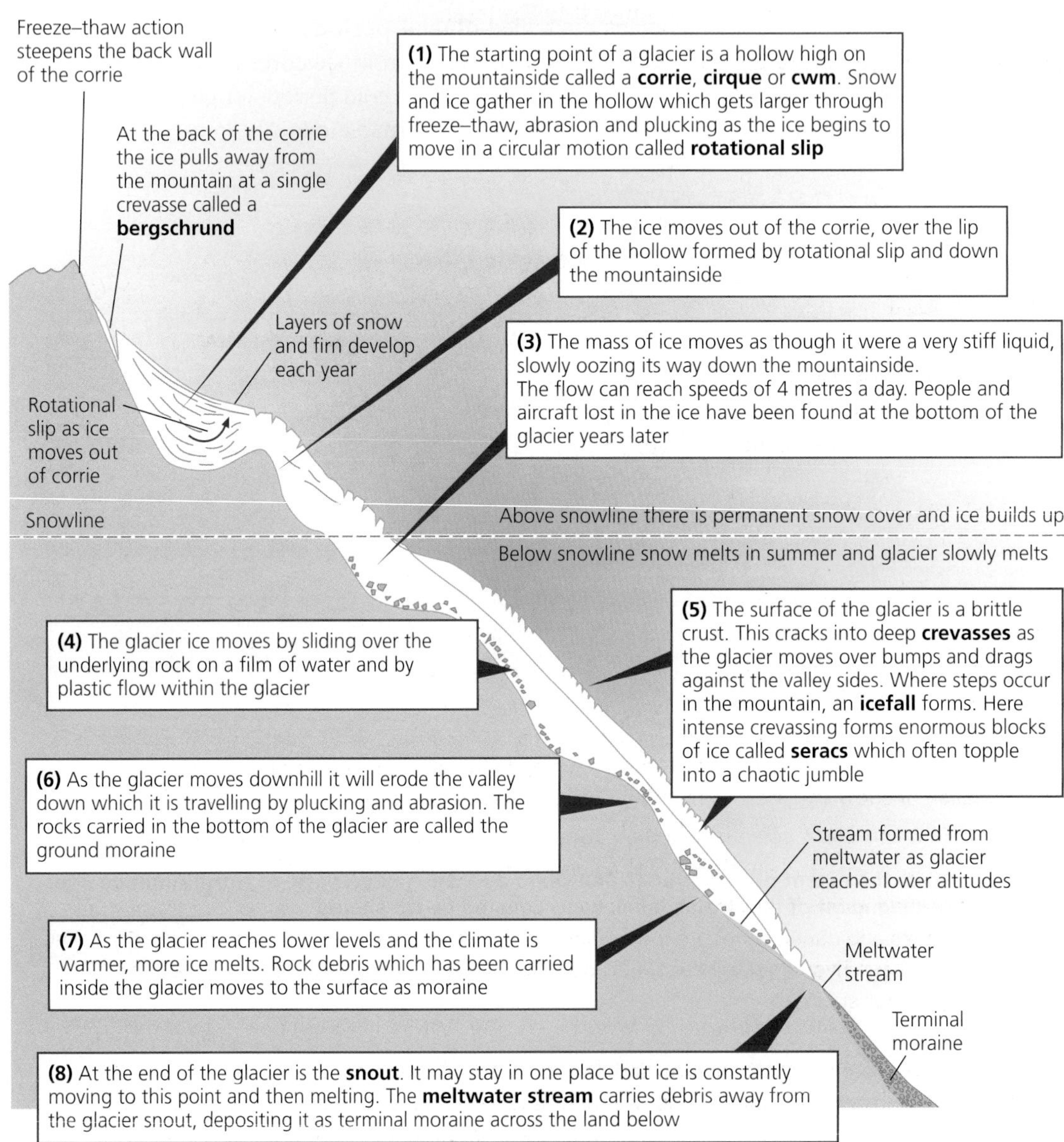

Freeze–thaw action steepens the back wall of the corrie
(1) The starting point of a glacier is a hollow high on the mountainside called a corrie, cirque or cwm. Snow and ice gather in the hollow which gets larger through freeze–thaw, abrasion and plucking as the ice begins to move in a circular motion called rotational slip
At the back of the corrie the ice pulls away from the mountain at a single crevasse called a bergschrund
(2) The ice moves out of the corrie, over the lip of the hollow formed by rotational slip and down the mountainside
Layers of snow and firn develop each year
(3) The mass of ice moves as though it were a very stiff liquid, slowly oozing its way down the mountainside. The flow can reach speeds of 4 metres a day. People and aircraft lost in the ice have been found at the bottom of the glacier years later
Rotational slip as ice moves out of corrie
Snowline
Above snowline there is permanent snow cover and ice builds up
Below snowline snow melts in summer and glacier slowly melts
(5) The surface of the glacier is a brittle crust. This cracks into deep crevasses as the glacier moves over bumps and drags against the valley sides. Where steps occur in the mountain, an icefall forms. Here intense crevassing forms enormous blocks of ice called seracs which often topple into a chaotic jumble
(4) The glacier ice moves by sliding over the underlying rock on a film of water and by plastic flow within the glacier
(6) As the glacier moves downhill it will erode the valley down which it is travelling by plucking and abrasion. The rocks carried in the bottom of the glacier are called the ground moraine
Stream formed from meltwater as glacier reaches lower altitudes
(7) As the glacier reaches lower levels and the climate is warmer, more ice melts. Rock debris which has been carried inside the glacier moves to the surface as moraine
Meltwater stream
Terminal moraine
(8) At the end of the glacier is the snout. It may stay in one place but ice is constantly moving to this point and then melting. The meltwater stream carries debris away from the glacier snout, depositing it as terminal moraine across the land below

The glacial budget

A glacier is a system of inputs and outputs. The main input or **accumulation** is snow that becomes compacted and turns into ice. The main output or **ablation** is melting ice that becomes meltwater. Most ice melts from the snout and from the surface of the glacier. Occasionally chunks of ice break away.

The **glacial budget** is the balance between accumulation and ablation. If the inputs are greater than the outputs the glacier will increase in size – usually by advancing further down the valley. If the outputs are greater than the inputs then the glacier will retreat. In winter, the glacier is likely to advance but in summer it will retreat.

Case study: *a glacier*

Khumbu Glacier, Nepal

The Khumbu Glacier starts in the Western cwm on the slopes of Mount Everest. It is the largest glacier in Nepal and is about 4 km (2.5 miles) long.

Everest base camp is on the glacier and climbers attempting to reach the Everest summit must first cross the dangerous Khumbu icefall. Ropes and ladders are used to cross the ice fall but even so many people have died either falling into crevasses beneath the snow or been crushed by ice falls.

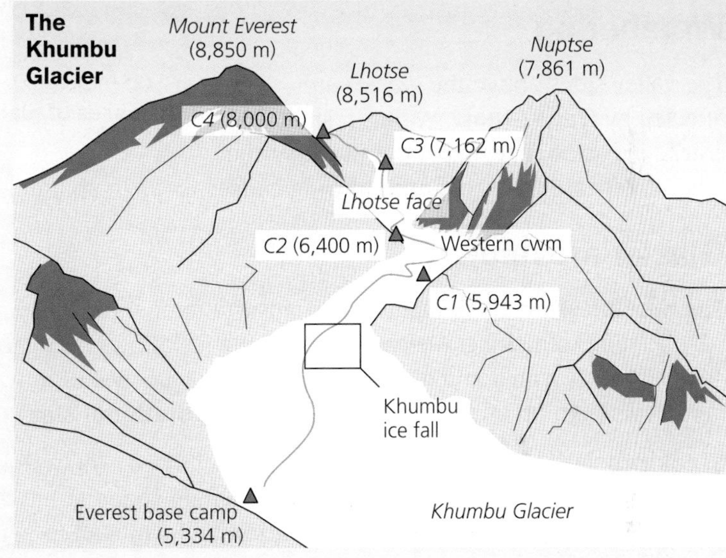

Since 1980, glaciers all over the world have been retreating and since 1995 this process has speeded up. Glaciers retreat if the amount of melting in the summer is greater than the accumulation of snow and ice in winter. It is thought that climate change is responsible for this.

Between 1976 and 2007, scientists studied 15 glaciers in the Everest region and found that all were retreating. They found the Khumbu Glacier is retreating by approximately 18 m a year.

The effect of melting glaciers is that rivers flowing from the glaciers through Nepal and India can burst their banks and cause widespread flooding. However, in the long term, as glaciers get smaller, water levels in these rivers may fall causing problems for people who depend on the rivers for irrigation and water supply.

In 2010, five cameras were placed in the Mount Everest region to monitor melting glaciers. The cameras take a photo every 30 minutes and after 6 months the photos are used to make a short video using time-lapse techniques. This two-year study will help scientists investigating glacial retreat and will also provide the Nepali government with important data so they can plan for the future.

Glacial processes

Ice is a powerful force in shaping the land through the following processes.

Erosion

As the glacier moves downhill it smoothes and deepens the valley by **abrasion** and **plucking**.

Ice flow

Sharp angular rocks embedded in ice

Rock

Blocks of rock pulled away by ice = **plucking**

Rocks scratch and scrape surface = **abrasion**

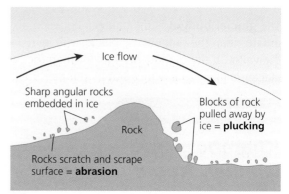

Processes of glacial erosion

Weathering

The valley sides above the glacier are affected by **freeze–thaw**. Water in small cracks in rock freeze, expand and widen the cracks. Repeated freezing and thawing causes rocks to shatter.

Key words

- abrasion
- plucking
- freeze–thaw
- basal flow
- rotational slip
- internal deformation
- moraine
- rock flour

Glacial movement

The huge pressure and weight of ice causes a temperature rise at the base leading to melting. Meltwater under the ice allows the glacier to slide slowly downhill and is called **basal flow**. This type of movement in hollows high up on the mountain results in **rotational slip**.

In colder polar regions the glacier is frozen to its bed. The weight of the ice causes individual ice crystals to change shape (a bit like plastic). This is called **internal deformation** and can lead to very slow downhill movement of the glacier.

Transportation

The glacier carries a large amount of rock debris known as **moraine**. This varies in size from rocks as large as a house to powder called **rock flour**. Most of the load is near the base of the glacier.

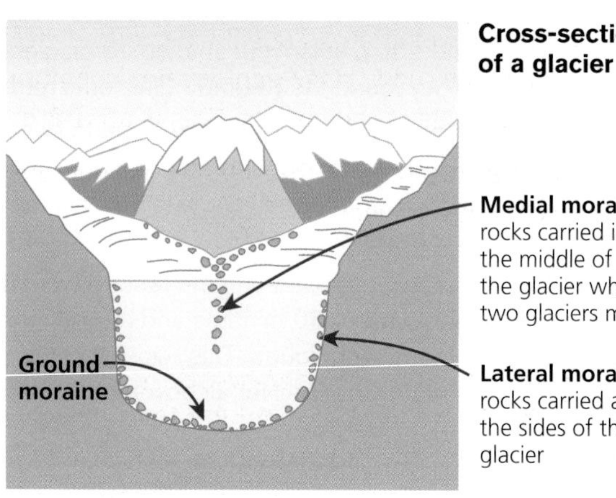

Cross-section of a glacier

Deposition

At the snout the ice melts and moraine is dumped in piles or washed away by the meltwater to be deposited over lowland.

Glaciated landforms

Landforms resulting from erosion

Ice is a powerful agent of erosion; the effects of glacial erosion are greatest in mountain regions where ice has been present for a long time.

Corries, arêtes and pyramidal peaks

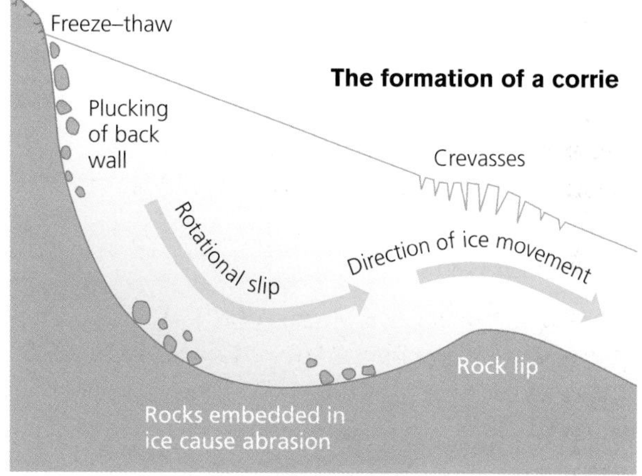

A **corrie** is a deep armchair-shaped hollow high up in the mountains. Corries begin to form when snow accumulates in a small hollow on the mountainside. The snow is compacted and turns into ice. As more and more ice accumulates it starts to move, due to the pull of gravity. This erodes and deepens the hollow as the ice freezes onto the rock, pulling it away from the wall of the corrie. This plucking process forms the steep and craggy back wall of the corrie. At the top of the back wall freeze–thaw weathering occurs as water seeps into the cracks and then re-freezes and expands, levering off pieces of rock. Rocks become embedded in the ice and help in the process of abrasion, scraping and eroding the surrounding rock. When ice melts at the end of the ice age it leaves a deep hollow. **Corrie lakes** (tarns) often develop in the base of the hollow.

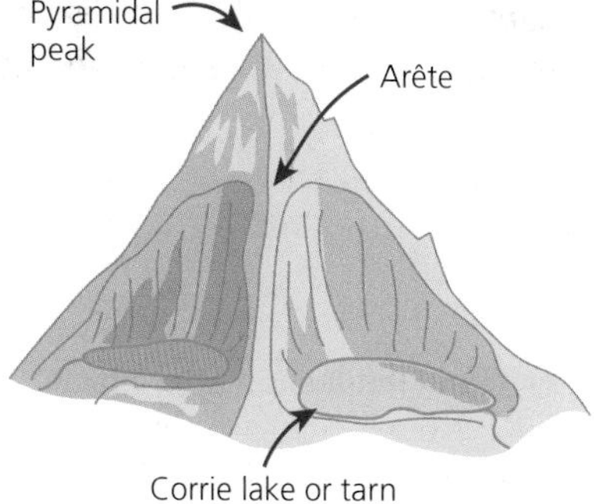

If two corries develop on adjacent sides of the mountain a very steep knife-edge ridge can develop separating the two back walls of the corries. This is an **arête**.

If three corries develop on the sides of a mountain a **pyramidal peak** develops.

Ribbon lakes

Long thin lakes, known as **ribbon lakes**, may develop on the floor of a **U-shaped valley**. This may be due to over-deepening of the valley floor, e.g. as a result of less resistant bed rock, or because two glaciers merged causing more erosion. Lakes may also be caused because the valley floor is blocked by moraine, which acts as a dam creating a lake upstream

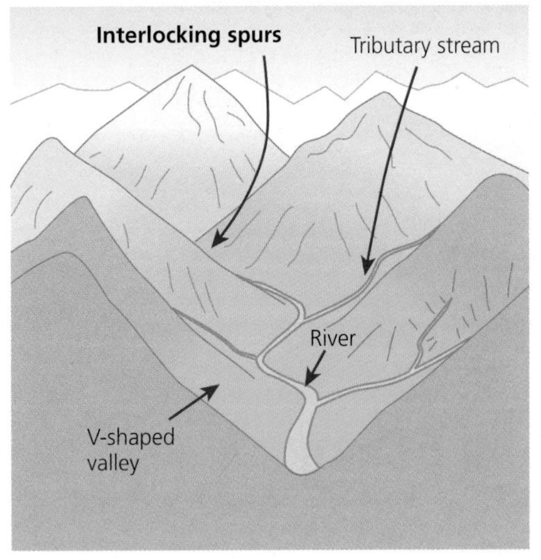

A highland landscape before glaciation

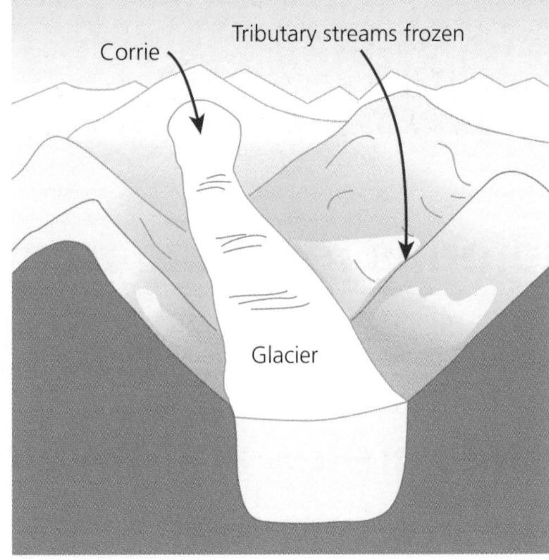

A highland landscape during glaciation

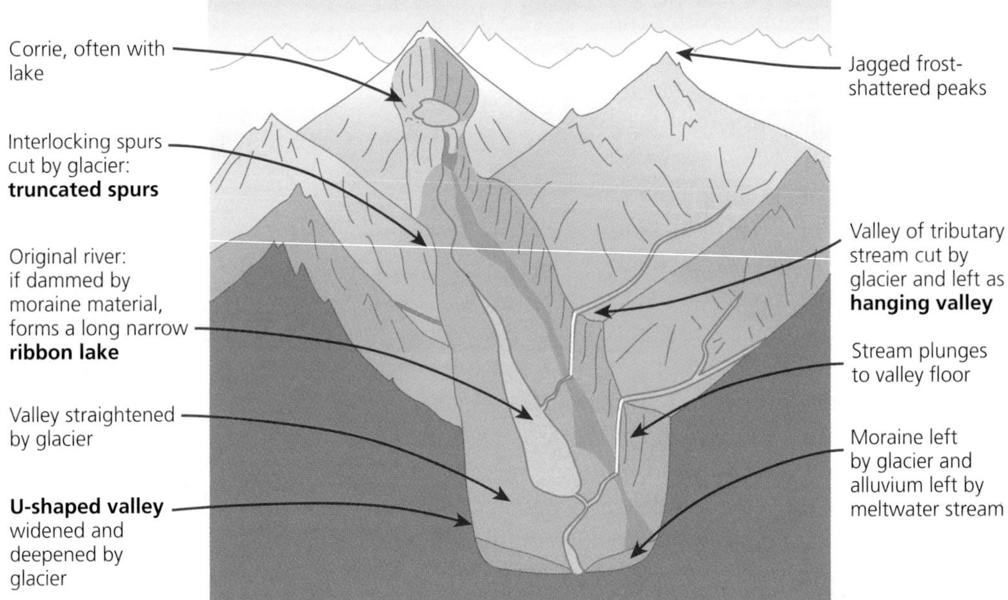

A highland landscape after glaciation

of the deposit, these will be much shallower lakes than those in areas of over-deepening.

A valley which has been subject to glaciation is called a **glacial trough** but is usually referred to as a U-shaped valley. These valleys can be several hundred metres deep with very steep, almost vertical sides. Waterfalls may flow over the valley sides from **hanging valleys** above – these develop where former tributary valleys meet the main valley. These valleys were once occupied by smaller tributary glaciers that did not erode down to the same level as the main glacier.

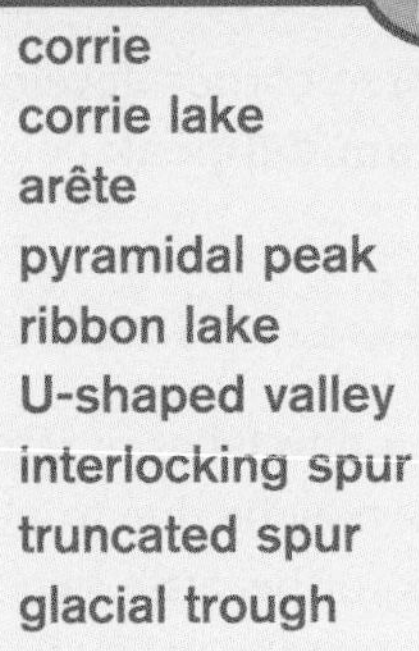

Key words

- corrie
- corrie lake
- arête
- pyramidal peak
- ribbon lake
- U-shaped valley
- interlocking spur
- truncated spur
- glacial trough

Landforms resulting from glacial transport and deposition

Moraine is a general name for material eroded from the sides and floor of the valley and carried by the glacier down the mountains. It is usually angular and can be large or small.

Lateral moraine forms at the edges of the glacier as material falls from the valley sides and either lies on top of the ice or becomes embedded in the ice. If two glaciers join then the two lateral moraines merge to form **medial moraine** running down the centre of the glacier. After the ice has melted, lateral and medial moraine form uneven hummocky deposits on the valley floor often in a linear shape.

Most material piles up at the snout of the glacier and this is called **terminal moraine**. After the glacier has retreated this may form a high ridge across the valley floor. It marks the furthest extent of the glacier.

Glaciers also transport a great deal of material underneath the glacier, this is called **ground moraine**. When the ice melts it is left behind – often as an uneven hummocky deposit.

The general name given to material deposited by glaciers is **boulder clay**.

The material in meltwater streams, deposited by water and not by ice, is usually sand and gravels. These can be carried over a wide area and form an **outwash plain**.

Drumlins are long, low hills deposited in lowland areas, shaped when seen from above like the back of a spoon with one blunt end and one tapered end. Drumlins are made of boulder clay (morainic material) and are usually about 10–40 m high and 100 m or more in length. They usually form in groups and their shape means they are said to look like a basket of eggs.

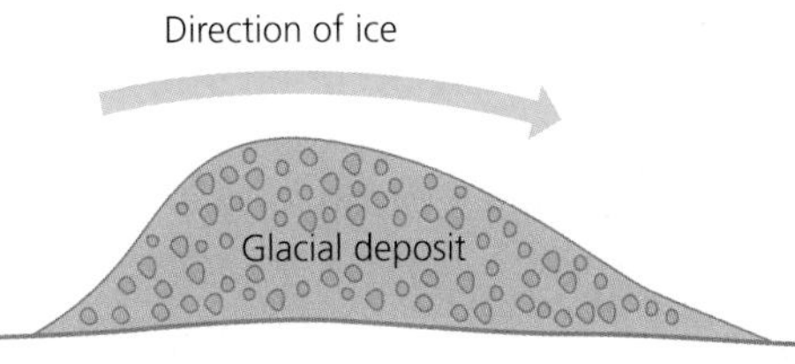

A drumlin

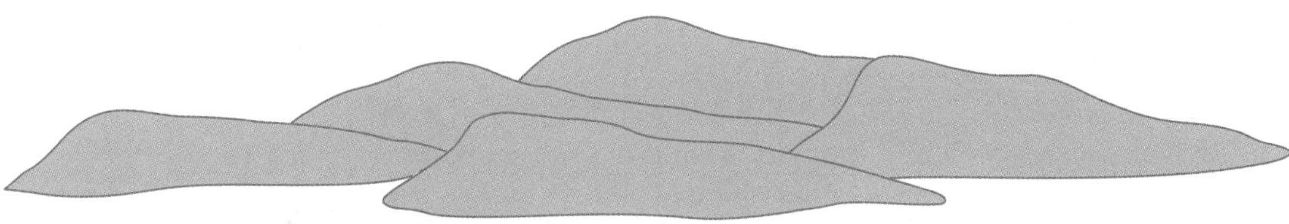

A group of drumlins — 'basket of eggs'

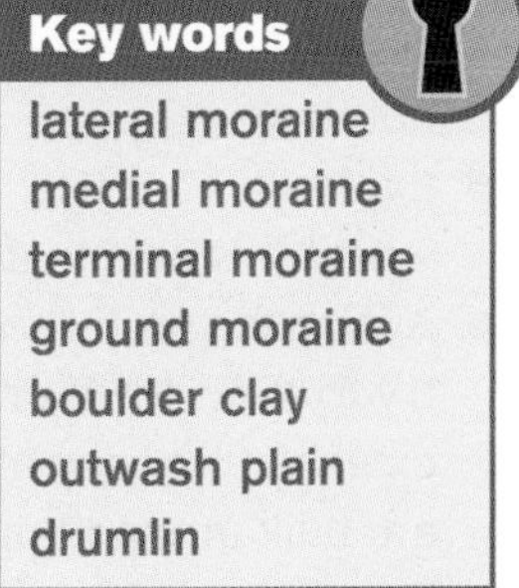

Key words

lateral moraine
medial moraine
terminal moraine
ground moraine
boulder clay
outwash plain
drumlin

Avalanche hazards

Every year in France alone, 25 to 30 people are killed by avalanches. The victims are usually skiing off piste (away from the marked ski runs) and six out of ten people buried in an avalanche die. Asphyxiation or hypothermia will kill most people within 30 minutes so survival depends on a quick response from rescuers. However, the average time for rescuers to reach an avalanche is 45 minutes.

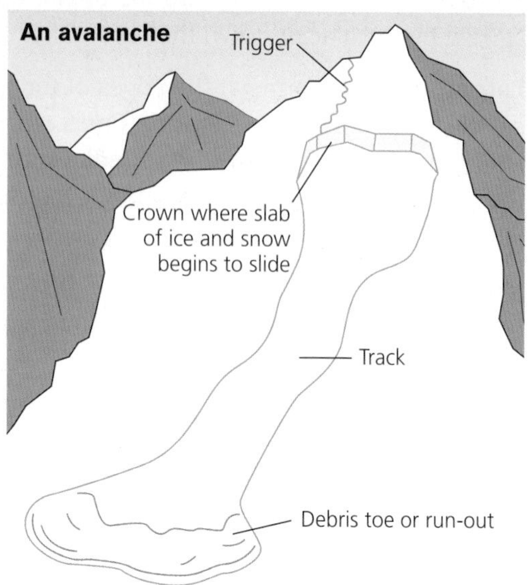

An avalanche is a mass of snow, ice and rocks moving very rapidly down a hillside at speeds up to 300 km/h. Avalanches have immense power and can cause a great deal of damage to people and property.

Causes of avalanches

Avalanches usually occur on steep slopes and can be triggered by a number of related factors including heavy snowfall, which adds weight to earlier snowfalls, or a sudden rise in temperature when snow starts to melt. Other causes can be:

- people's actions may trigger avalanches
- removal of trees for ski developments makes it easier for avalanches to move downhill
- skiing in areas of fresh snow

Avalanche protection

It is not possible to stop avalanches but it is possible to protect people and places and to warn them if avalanches are likely. This has become vitally important in ski resorts in the Alps and other popular winter sports destinations.

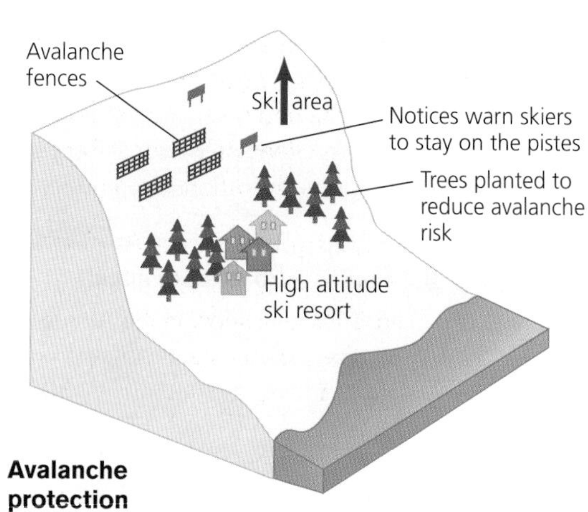

Avalanche protection

- Resorts such as Les Arcs have clear warning systems for skiers and notices telling people not to ski off piste.
- Explosives are used to trigger small avalanches where it is safe to do so.
- Avalanche protection structures are built to try to slow down or divert avalanches. For example, wooden avalanche fences, steel barriers and snow nets. Sometimes snow sheds are built around important buildings or structures that might be at risk.
- Trees are planted in areas up to the tree line to provide natural protection.

Case study: *an Alpine area for winter sports*

Tourism in Les Arcs in the French Alps

The Alps attract over 120 million visitors a year. Les Arcs in the Tarentaise Valley is one of France's largest and newest ski resorts, opened in the 1960s. Since 2003, it has been linked to La Plagne by an express cable car creating the largest ski area in Europe, Paradiski.

Les Arcs is between 1,600 and 2,000 m above sea level, with ski lifts climbing to over 3,000 m. Snow is guaranteed from December until April and glaciers provide summer skiing.

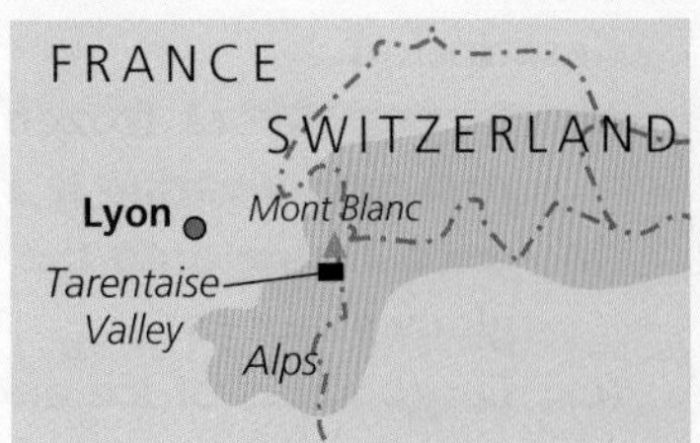

Location of the Tarentaise valley

Socio-economic impacts

- Hundreds of jobs in hotels, shops and restaurants. Less employment in summer when fewer visitors.
- The local economy has benefited — employment means local people have money to spend in local shops and services.
- Better access for local people although traffic congestion and noise can be a problem.
- Young people more likely to stay in area to find work rather than migrate to large towns.

Environmental impacts

- Building ski resorts — buildings, roads and ski lifts — makes the area look unattractive.
- The huge increase in traffic can cause noise and congestion especially at weekends.
- The main ski runs or pistes are damaging the fragile vegetation below. In summer these form wide scars.
- The fragile alpine ecosystem is threatened; some wildlife is at risk of losing habitat, e.g. eagles and lynx.
- Trees felled to make space for buildings, roads and ski runs, can lead to soil erosion and avalanches.
- Artificial snow used in the spring on the lower slopes requires millions of litres of water so large reservoirs have had to be built.

Management

- Les Arcs was built as a series of resorts so that visitors could easily access ski slopes.
- Electricity is generated by hydroelectric power stations, a renewable source of energy.
- Roads in resorts are pedestrianised; visitors must leave their cars at the edge of the villages.
- Warning systems and protection structures have been set up to protect people from avalanches.
- The area is in the Vanoise National Park so there are restrictions on how it is developed.
- New outdoor activities attract summer tourists so jobs in tourism are no longer seasonal.

What is the impact of climate change?

Some of the older resorts at low altitudes are beginning to face difficulties with less snow. By 2030:

- there could 30% less snow in the Alps as a result of climate change
- extreme weather conditions including very heavy snow falls are more likely
- the snowline in the Alps could rise by 300 m leaving 40% of resorts without enough snow

Is tourism in the Alps sustainable?

In the long term, tourism may damage the environment and visitors could choose other resorts. The Alpine landscape must be protected to maintain economic prosperity by:

- investing in snow canons to make artificial snow to keep the lower slopes covered for skiing
- developing alternative activities for both summer and winter to attract different types of tourists
- building resorts higher with access to glacier skiing, but this threatens the fragile environment

Test yourself

1 **Match the correct word from the list on the right to each description.**

(a) Pieces of rock frozen in the ice scraping at the surrounding rock and wearing it away.

(b) Ice freezing onto the rock and pulling it away when the glacier moves.

(c) Large areas of ice and snow.

(d) A hollow on the mountainside left by ice.

(e) A single crevasse at the back of a corrie.

(f) Enormous blocks of ice caused by intense crevassing.

(g) A moraine formed at the side of a glacier.

(h) A moraine formed in the centre of a glacier when two glaciers meet.

(i) The lowest point of a glacier.

(j) A knife-edged ridge between two corries.

(k) A valley left high above the main valley floor.

(l) A long, low hill of deposited material found in lowland areas.

hanging valley
lateral moraine
plucking
snout
ice field
bergschrund
abrasion
medial moraine
drumlin
seracs
corrie
arête

Exam tip

Don't forget to use your case study knowledge to give examples of the points you make.

Examination question

Foundation tier:

(a) (i) Name an Alpine area you have studied and describe the social, economic and environmental impacts tourists have in this area. *(6 marks)*

(ii) Describe two ways in which the area is being protected. *(2 marks)*

Higher tier:

(b) Using a case study of an Alpine area discuss the impacts tourism has on the area and the ways in which the area is being managed for the future. *(8 marks)*

Waves and erosion

Two basic factors affect the nature of the coastline: the waves and the type of rock.

Waves

The movement of water particles in a wave is circular.

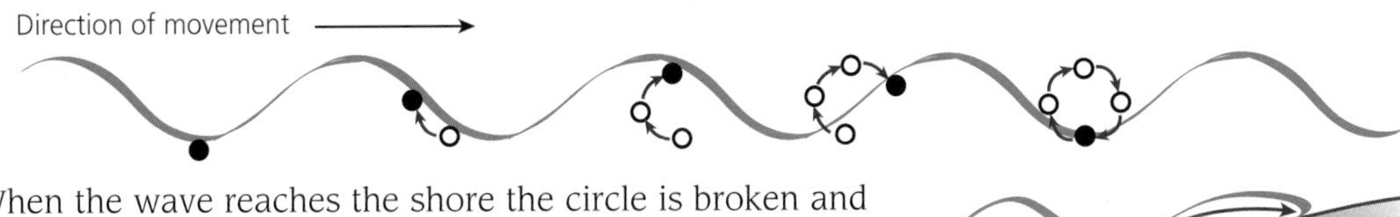

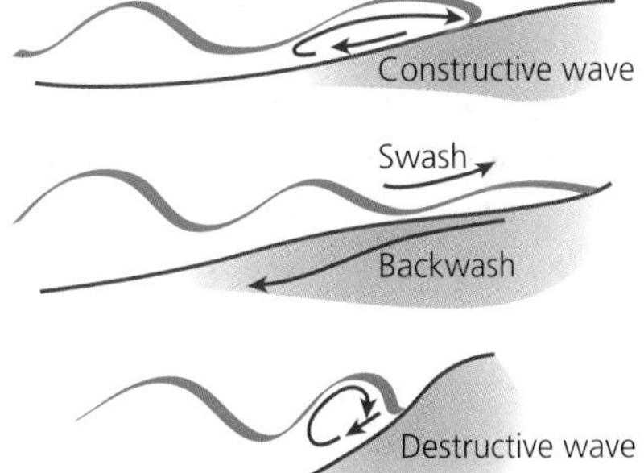

When the wave reaches the shore the circle is broken and the wave spills forward – it breaks. If the slope of the shore is shallow, the wave spills forward for a long distance and is called a **constructive wave** because it pushes material onto the beach.

As the wave breaks, it swills up the beach. This is known as **swash**. It then runs straight back down the beach – known as **backwash**. If the slope of the shore is steep, the wave plunges down and hits the shore with great force. It is called a **destructive wave** because it erodes the coast.

Processes of erosion by waves

Waves erode the coastline in four main ways:

1 **Abrasion** or **corrasion** The sea hurls pebbles and sand against the base of the cliff, chipping and grinding it down.
2 **Hydraulic action** Powerful waves lash the cliffs, forcing air into tiny cracks. The pressure of the compressed air weakens the rock and forces it to break up.
3 **Corrosion** The sea water may react with chemicals and minerals in some rocks and they can be dissolved.
4 **Attrition** The rocks and stones that the sea erodes from the cliffs are rounded and broken down as they bump against each other and they are thrown against the cliff.

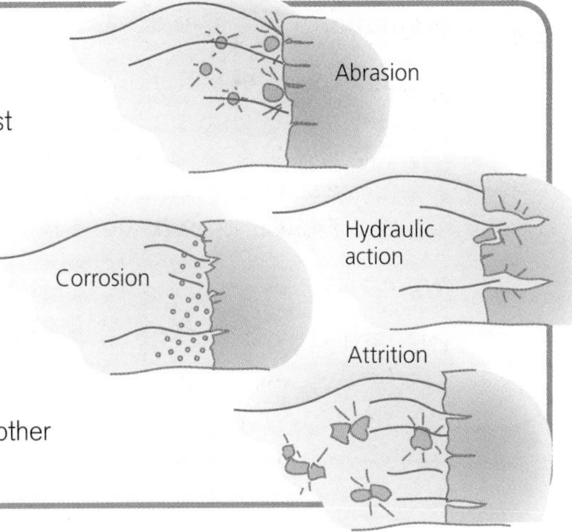

Weathering

The cliff face is likely to be eroded by weathering.

Mechanical weathering

In resistant rocks such as limestone and chalk, water can get into cracks in the rock and widen them through **freeze–thaw**.

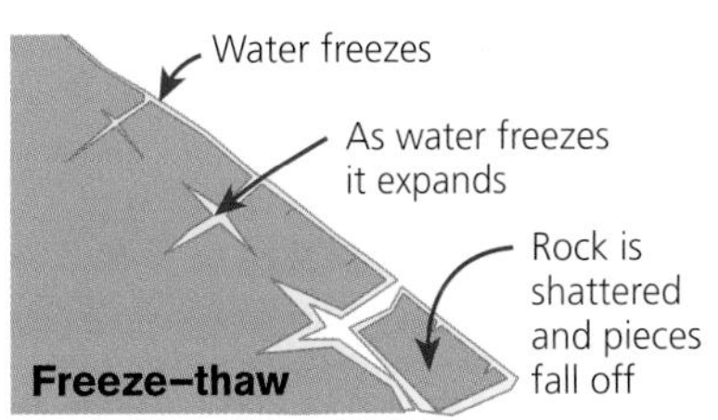

Chemical weathering

Some minerals and rocks dissolve in rainwater, which is slightly acidic. Limestone and granite are vulnerable to chemical weathering.

Mass movement

Any downhill movement of material is called **mass movement**, e.g. rock falls caused by freeze–thaw in resistant rock. Mass movement also occurs in cliffs formed in less resistant rock such as sands and clay, which are more easily eroded. These cliffs are eroded at the foot by the sea, but the cliffs themselves can become waterlogged by rainfall causing them to become unstable and leading to land slips also called rotational slumping.

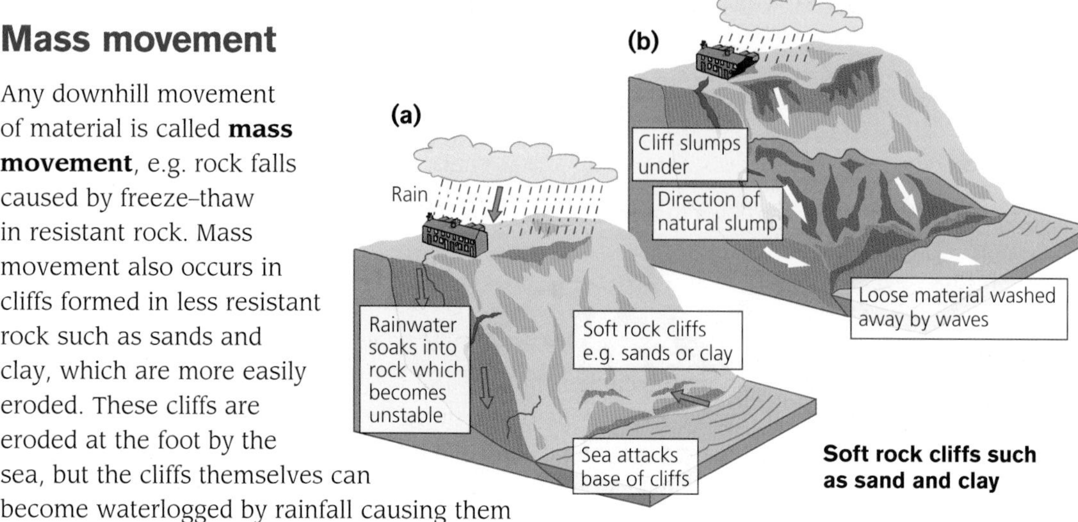

Soft rock cliffs such as sand and clay

Processes of transportation

The largest materials such as boulders are rolled along the sea floor by waves (**traction**); smaller materials such as pebbles are bounced (**saltation**); finer material such as sand is carried in **suspension** and soluble material such as limestone is dissolved and carried in **solution**.

The movement of material by waves along the coast is called **longshore drift**. The movement of sediment depends on the direction of the prevailing wind.

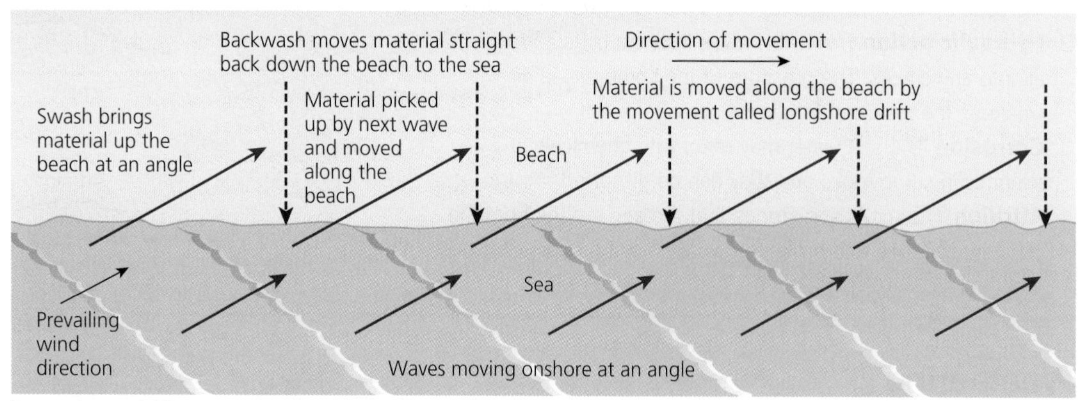

Longshore drift

Deposition

Sediment is deposited where the sea no longer has enough energy for it to be transported. This happens in areas of calm water, e.g. in a bay.

Landforms resulting from erosion

Cliffs and wave-cut platforms

Cliffs formed from hard and resistant rocks such as chalk or limestone are eroded slowly. The cliffs are often high and almost vertical and erosion by the sea at the base of the cliff can cause rock falls.

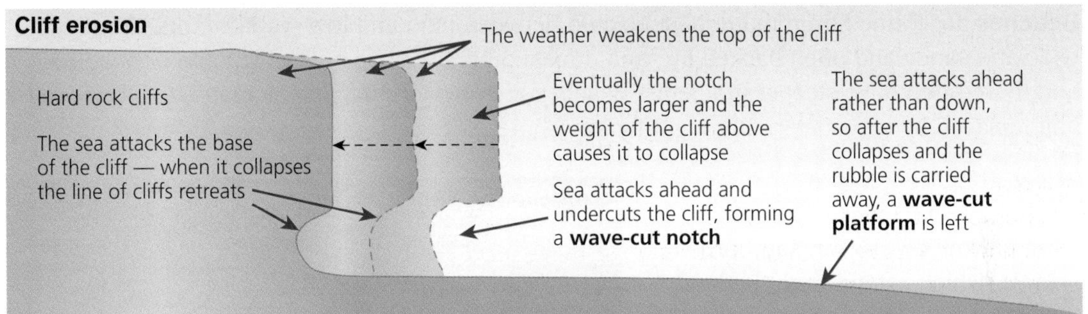
Cliff erosion
The weather weakens the top of the cliff
Hard rock cliffs
The sea attacks the base of the cliff — when it collapses the line of cliffs retreats
Eventually the notch becomes larger and the weight of the cliff above causes it to collapse
Sea attacks ahead and undercuts the cliff, forming a **wave-cut notch**
The sea attacks ahead rather than down, so after the cliff collapses and the rubble is carried away, a **wave-cut platform** is left

Headlands and bays

Where rocks of different resistance lie at right angles to the coast the weaker rocks are eroded more quickly forming **bays**. More resistant rock sticks out to form **headlands**, which are steep rocky promontories jutting out into the sea.

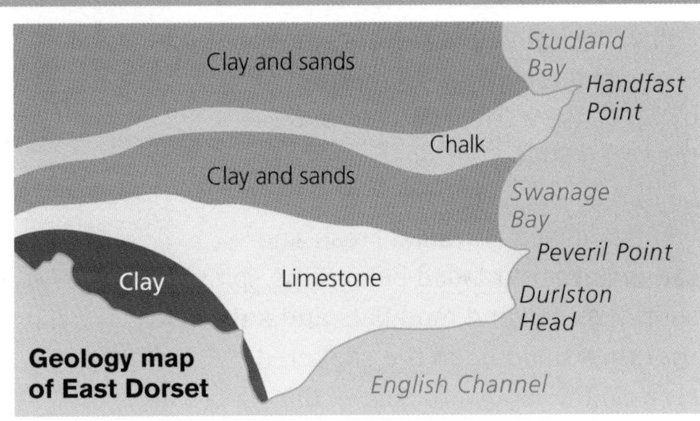
Clay and sands
Studland Bay
Handfast Point
Chalk
Clay and sands
Swanage Bay
Peveril Point
Clay
Limestone
Durlston Head
Geology map of East Dorset
English Channel

Caves, arches and stacks

Where a headland develops the sea can attack from three sides.

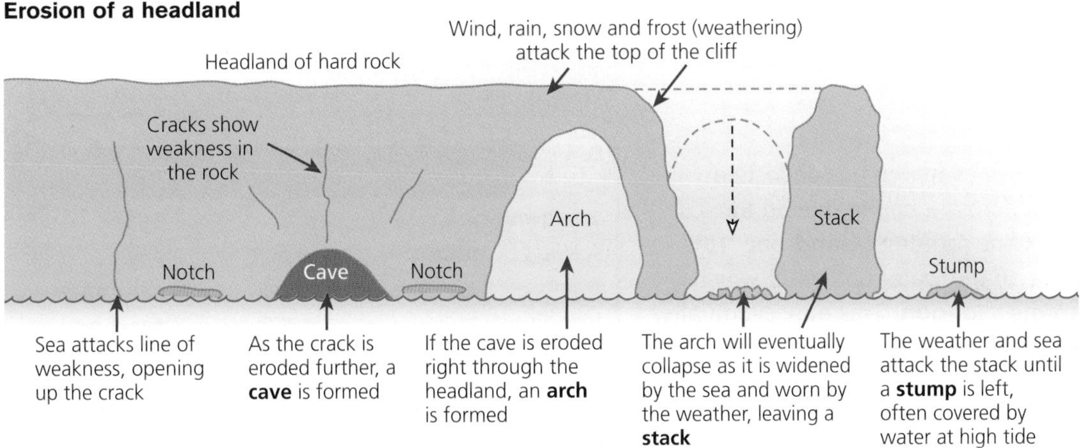
Erosion of a headland
Wind, rain, snow and frost (weathering) attack the top of the cliff
Headland of hard rock
Cracks show weakness in the rock
Arch
Stack
Notch
Cave
Notch
Stump
Sea attacks line of weakness, opening up the crack
As the crack is eroded further, a **cave** is formed
If the cave is eroded right through the headland, an **arch** is formed
The arch will eventually collapse as it is widened by the sea and worn by the weather, leaving a **stack**
The weather and sea attack the stack until a **stump** is left, often covered by water at high tide

Landforms of deposition

The coastline is constantly changing because erosion, transport and deposition is occurring. Waves transport sediment along the coast by **longshore drift** and this may result in new land being created through deposition.

Beaches

Beaches are formed from material deposited between high and low water marks. Some are wide and sandy and often backed by sand dunes; others may be built of shingle or pebbles and these often have steeper gradients. Beaches are always changing as material is removed and replaced by new material brought by longshore drift.

Spits

Long narrow deposits of sand or shingle, which extend from the land out into the sea, are called **spits**. Spits develop where the coastline changes direction or is interrupted by an estuary. The seaward end of the spit is in deeper water and may be curved because of the effect of winds and sea currents. Fresh and salt water are trapped behind the spit as it develops and mud flats and **salt marshes** build up in the sheltered area on the landward side of the spit. A well-known example of a spit is Spurn Head, which extends for 4.8 km (3 miles) across the Humber Estuary.

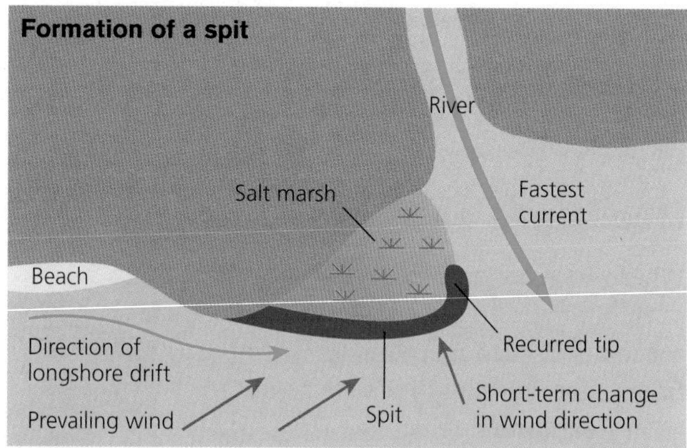

Bars

Sometimes material is deposited right across a bay or inlet in the coastline forming a **sand bar**. This cuts off the water behind it, which forms a **lagoon**. This may eventually disappear and become colonised by plants.

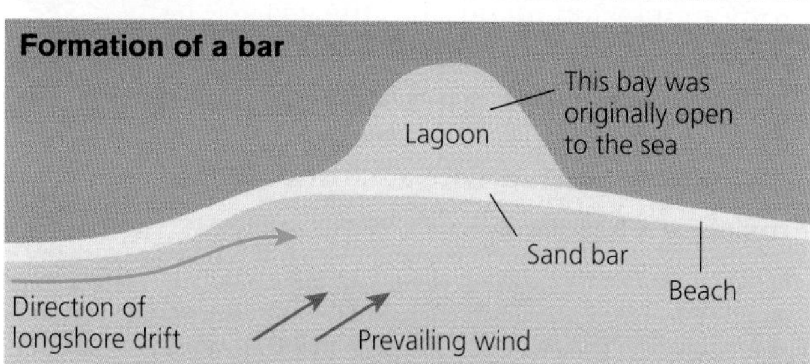

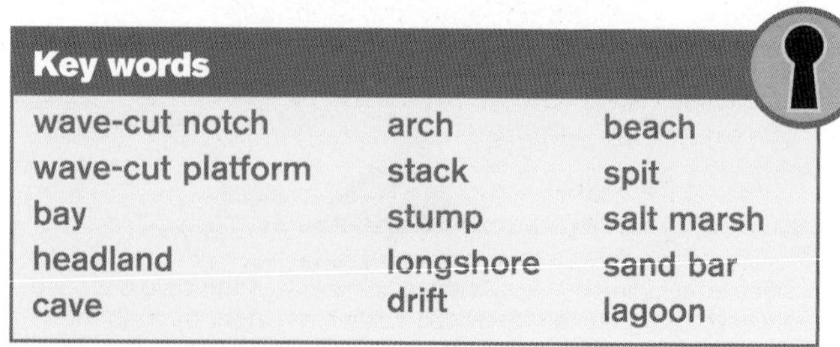

Key words

wave-cut notch	arch	beach
wave-cut platform	stack	spit
bay	stump	salt marsh
headland	longshore drift	sand bar
cave		lagoon

Effects of rising sea levels

Global sea levels have risen on average about 3 mm a year in the last 15 years. By 2100, sea levels may have risen a further 30–40 cm. This is enough to submerge low lying places.

Causes of rising sea levels

- Thermal expansion of seawater caused by increasing air temperatures.
- Melting continental ice sheets, for example from Greenland, which will add water to the sea.

Case study: *impact of coastal flooding*

London and the Thames Estuary

London and the Thames Estuary are liable to flood if high tides combine with a storm surge.

Impacts of flooding

Flood risk zone

Social and economic impacts

- Over 1.25 million people live or work in the flood risk zone.
- 400 schools, 16 hospitals, eight power stations, City Airport, 30 mainline railway stations and 38 underground stations, including most of the central part of the underground network, are at risk of flooding.
- 120,000 new houses are planned in the flood risk zone.
- Damage to buildings and infrastructure would have serious economic impacts.

Political impacts

- Houses of Parliament and other government buildings could be flooded.
- Cost of protecting London.

Environmental impacts

The area at risk of flooding contains important habitats for fish, invertebrates and birds. Floods might result in untreated sewage and industrial waste getting into the river.

Thames flood defences

The Thames Barrier opened in 1984 and is the most important part of London's flood defences. The Barrier is closed about five times a year. There are also eight smaller barriers and several miles of floodwalls and embankments to protect London.

The Environment Agency plans to:

- maintain and modify the Thames Barrier so it can provide protection until 2070
- raise the height of flood walls and embankments alongside the river
- construct another barrier at Long Reach, near Dartford, to be ready by 2075 if needed; this will not be decided until 2050

Coastal management

Some sections of coastline need to be protected to prevent rapid erosion from the sea, but it is too expensive to protect every stretch of vulnerable coastline. Protecting one area of coast can result in other areas being eroded more quickly or being more likely to flood.

Where the sea is eroding the coastline it can do so at an alarming rate. This may not be seen as an issue when farmland is involved, but if homes and towns are threatened then it is more serious. There are several common methods used to try and stop the erosion.

Soft engineering

These approaches are usually cheaper and do not damage the appearance of the coast. They are therefore a more sustainable approach to coastal protection. On the other hand they are usually not as effective as hard engineering methods.

Beach replenishment Sand is brought in to build up the beach either from further along the coast or from offshore. This looks completely natural and provides a beach to protect the coast and for visitors to enjoy. However, the sea will continue to erode the beach so replenishment has to be repeated every few years.

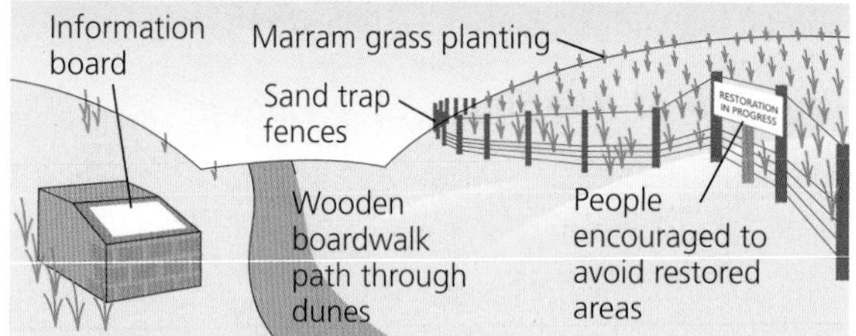

Dune management Sand dunes provide good natural protection for the coast. Dunes may be damaged by storms or by visitors walking through them to get to a beach. Dune management includes planting marram grass to stabilise the sand or fill gaps and by making wooden boardwalks for use as footpaths to reduce visitor impact.

Managed retreat

In some places the sea is being allowed to erode the coast and people and activities have to move away. This is clearly the cheapest solution but it is very disruptive for the people who live where land and buildings are likely to be lost. In many cases compensation is not paid so individuals can lose a great deal of money. It can be very stressful and disruptive so has important social costs. However, this approach will increase the amount of salt marshes, which help to stop future erosion and flooding and also provide a habitat for birds and wildlife.

Integrated coastal zone management (ICZM)

In many areas planners decide to combine hard engineering schemes with soft engineering and managed retreat. An integrated approach is used on the Holderness coast in Yorkshire (see page 76).

Hard engineering

These structures are very expensive to build and maintain and are only used when towns, villages or expensive installations are at risk and where the economic benefit is greater than the costs involved.

- **Groynes** At seaside resorts wooden walls or groynes are built across the beach to stop the sand being washed away by longshore drift. The beach material builds up at one side of the groyne. Trapping the material like this may cause problems elsewhere as it stops the material moving down the coast where, for instance, it may be building up and protecting the base of a cliff. New groynes are expensive and need to be maintained to stop the wood from rotting.
- **Sea walls** The most effective method of halting sea erosion. They are also the most expensive and cost about £500,000 per metre to build. Made of concrete, they are curved to deflect the power of the waves. But the sea can undermine them if the beach material in front of them is not maintained. Sea walls may by unsightly and also can restrict access to the beach.
- **Revetments** These are sloping wooden fences with an open structure of planks to break the force of the waves and trap beach material behind them, protecting the base of the cliffs. They are cheaper but not as effective as sea walls.
- **Gabions** Less expensive than a sea wall or a revetment. They are cages of boulders built up at the foot of the cliff or on a sea wall.
- **Rip rap** or **Rock armour** The cheapest method but still expensive. It entails placing piles of large boulders on the beach to protect the cliffs from the full force of the sea.
- **Off-shore breakwater** These are built on the sea bed a short distance from the coast and are usually made from rock or concrete. They are also very expensive to build but are effective because the waves break on the barrier before reaching the coast. This reduces wave energy and allows a beach to build up, which protects the cliffs.

Examples of hard engineering structures

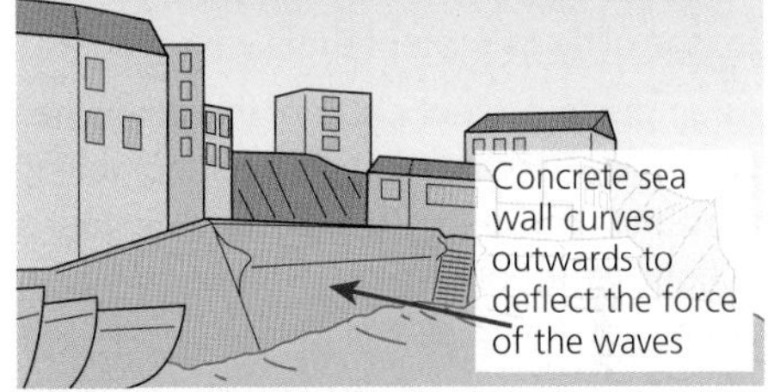

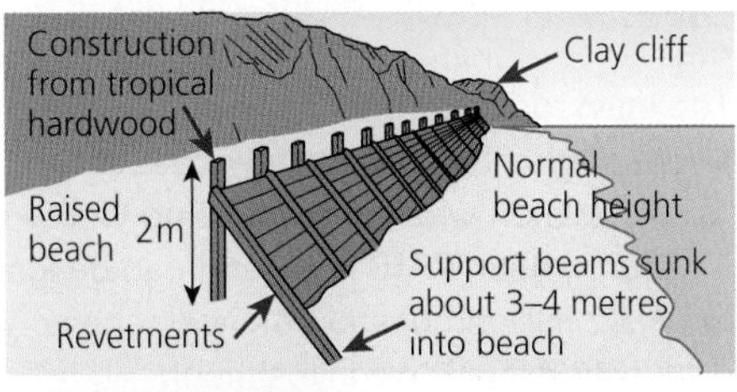

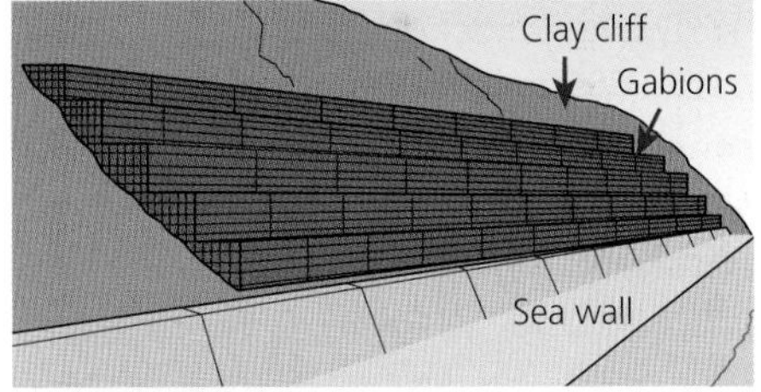

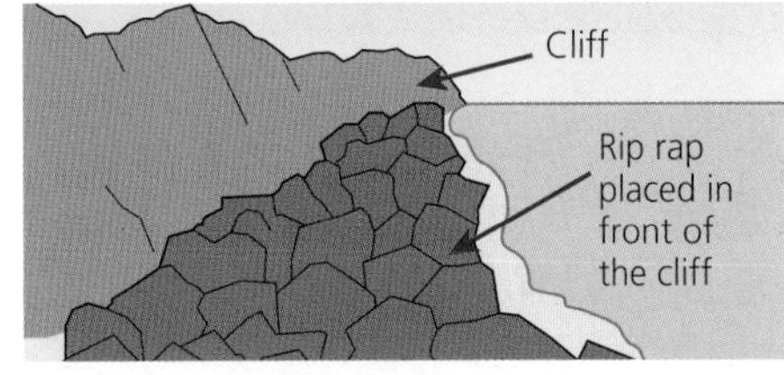

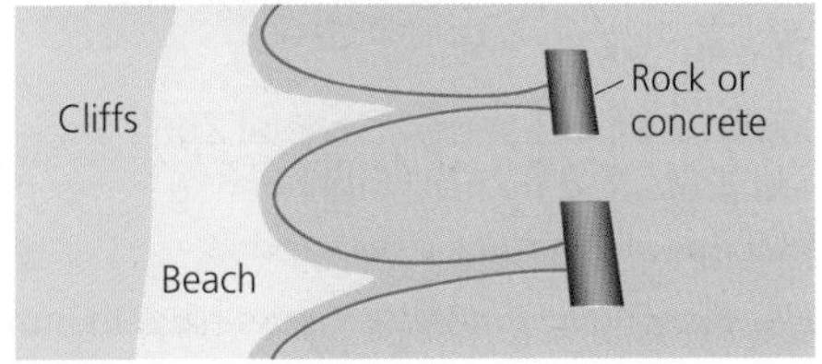

Case study: *an area coastal management*

Mappleton, Holderness

The coast between Flamborough Head and Spurn Head is formed of boulder clay and is being eroded. Four kilometres of land has been lost in the last 2,000 years.

The coast is being eroded rapidly because:

- Boulder clay is soft, unconsolidated rock and is easily eroded. The sea attacks the cliff foot, which is eroded by hydraulic action and abrasion. The cliff face is eroded by weathering which causes rotational slip.
- Longshore drift sweeps the eroded material south so beaches do not build up in front of the cliffs to protect them.

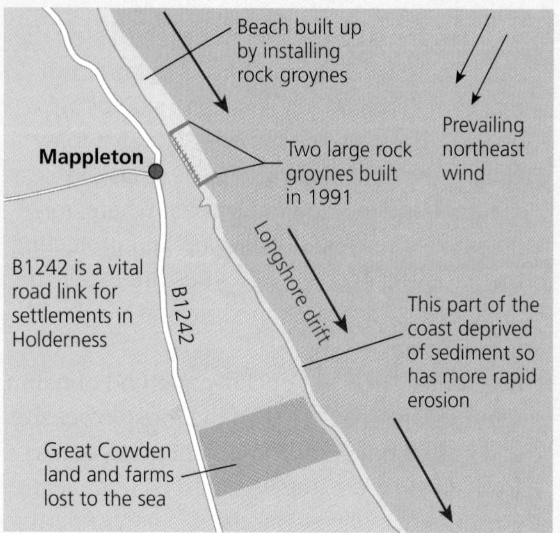

Attempts to protect parts of the coastline in the past have also been blamed for increasing erosion in places further along the coast so the actions of people also contribute to the rapid erosion of this coastline.

Mappleton is a small village with about 50 properties which lies 3 km south of Hornsea. The main road through the village (B1242) links all the settlements along the coast.

- Mappleton was at risk of being lost to the sea; on average 2 m of land is lost each year but in a stormy year, 7–10 m of land can be lost.
- A £2 million coastal defence scheme was approved in 1991. The risk of losing the vital road link justified this huge investment.
- Two large rock groynes were built and a rock revetment was made along the base of the cliff; 60,000 tonnes of granite blocks were brought from Norway.
- The groynes interrupt longshore drift and build up a beach.

Costs of the scheme

- It was very expensive – over £2 million.
- Erosion south of the groynes has increased significantly because sediment is trapped by the groynes so there is no beach to protect the cliffs to the south, which are now eroding faster than before.
- Land and buildings have been lost at Great Cowden. Farmers here had to pay for the demolition of their own farm buildings as the council refused to pay compensation.

Benefits of the scheme

- The cliffs at Mappleton are no longer at risk as a wide beach has built up between the two groynes and the revetment prevents cliff foot erosion.
- The village is protected and the road has not had to be rebuilt.

The future

In 2002 an Integrated Coastal Zone Management plan was produced for Holderness. Some places are protected by hard engineering but some are unprotected. This **managed retreat** means some places will eventually disappear.

People who live in these areas have to move when their land disappears. They receive no government help and even have to pay for the demolition costs of buildings themselves.

Salt marshes: a coastal habitat

Salt marshes develop in sheltered river estuaries or behind spits or bars where sediment is deposited. They form where salt water and fresh water meet and where there are no strong tides to wash sediment away. They are covered at high tide but exposed at low tide.

As sediment accumulates it forms mud flats. These are exposed at low tide and **colonised by plants** that can tolerate salt and also tolerate being submerged under water. These specially adapted plants may have thick, fleshy or hairy leaves to hold the moisture. Cordgrass is a pioneer species often first to colonise the mud; it has long roots to keep it in place. Cordgrass helps to trap more sediment allowing other **salt tolerant plants** to grow. Thus the mud flats become a salt marsh.

***Case study:** a coastal habitat*

Spurn Head salt marsh

Spurn Head is a narrow sand spit on the Yorkshire coast which extends across the Humber Estuary. It is 4.8 km long (3 miles) but in places less than 50 m wide. Mud flats and a salt marsh have developed on its western side sheltered by the spit.

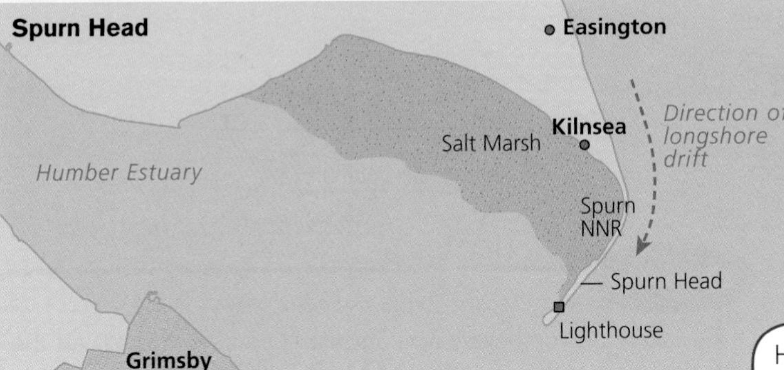

The environment

- The spit was formed by longshore drift moving sediment southwards along the Holderness coast and depositing it where the waters are sheltered by the Humber Estuary. The sand and shingle has built up over centuries to form the spit.
- Mud flats and salt marsh have developed in the shelter behind the spit.

Habitat and species

- Many salt tolerant plants such as cordgrass, glasswort and sea asters.
- Internationally important for birds, including wading birds and many migrating birds.

Strategies to protect the salt marsh

- Spurn Head has been designated a National Nature Reserve and a Site of Special Scientific Interest (SSSI). It is owned by the Yorkshire Wildlife Trust.
- Visitors are welcome to walk, fish or watch birds. Dogs are not allowed and cars must be left in car parks.
- Spurn Head is a remote place, which is being **managed sustainably**. In order to protect it for the future, access is limited and no development is allowed.
- In the future, the spit may be breached by high tides or rising sea levels and the spit and salt marshes will be lost to the sea.

Key words

salt marsh
plant colonisation
salt tolerant plants
sustainable management

Test yourself

1 Tick those statements which are correct and put a cross next to those which are incorrect.

(a) A constructive wave spills forward on a gently-sloping coast, pushing material onto the beach. ☐

(b) Soft rock such as clay is easily eroded by the weather and the sea, so it forms headlands. ☐

(c) Corrosion is the process by which the sea hurls pebbles at the cliffs, wearing them down. ☐

(d) Attrition is the process by which rocks from the cliffs are rounded and broken down. ☐

(e) A wave-cut platform is left at the base of a cliff after the cliff has collapsed. ☐

(f) When a headland arch collapses a notch is left standing offshore. ☐

(g) The zigzag movement of the swash and backwash causes longshore drift. ☐

(h) If material builds across the mouth of a bay due to longshore drift it forms a sand spit. ☐

(i) Groynes are built to prevent longshore drift washing away the sand at a seaside resort. ☐

(j) The Thames Barrier will soon be unable to protect London from floods. ☐

2 Label the diagrams below correctly and give each a title.

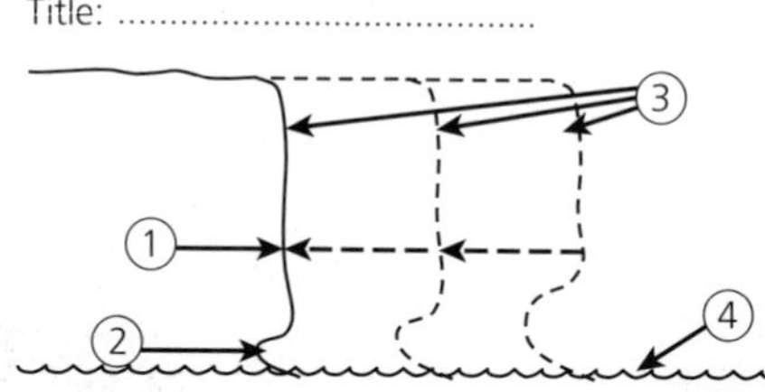

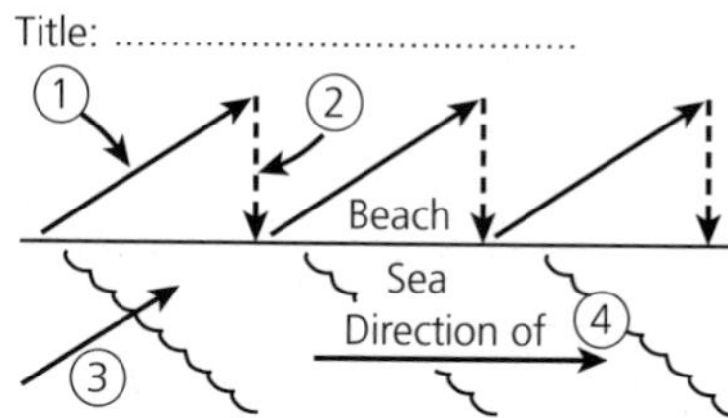

Exam tip

Make sure you have learnt and can use a case study for the final part of the question. The higher-tier question has three instructions or commands – can you find them?

Examination question

Foundation tier:

(a) (i) Name a coastal management scheme you have studied and describe two ways in which this coast is being protected from further erosion.

(ii) Explain some of the advantages and the disadvantages of the way the coast is being managed.

(8 marks)

Higher tier:

(b) Using a coastal management scheme you have studied, describe the ways in which it has been protected from erosion, and assess the costs and benefits of the strategies adopted.

(8 marks)

Unit 2
Human geography

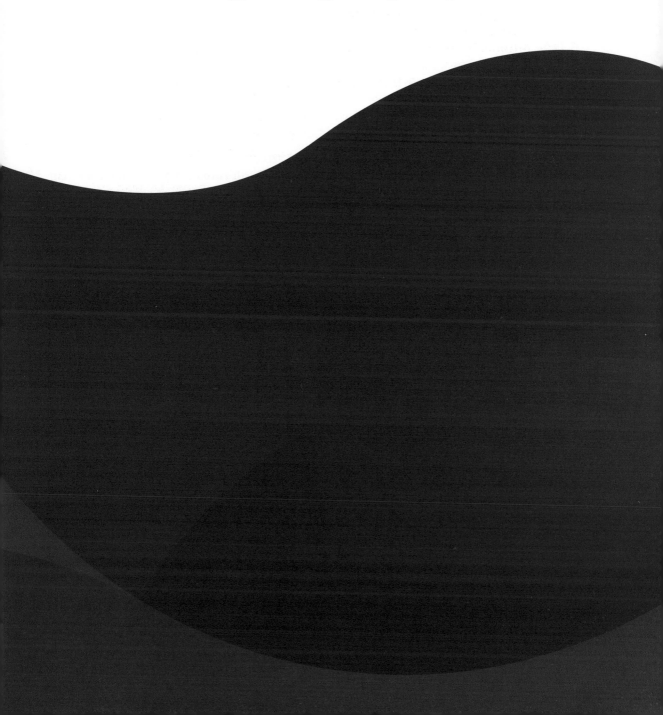

World population growth

- The world's population is growing rapidly.
- It took until 1804 to reach the first billion.
- The second billion took 123 years and the third billion took only 32 years.
- In 12 years, from 1987 to 1999, the population grew from 5 billion to 6 billion.
- It is expected to reach 7 billion in 2011.
- The latest prediction from **demographers** – the people who study population – is that the world population will level off at 8–10 billion around 2050.

Changing population

Ninety-five per cent of all population growth since 1950 has been in low-income countries, especially in the very poorest countries. This is because **birth rates** remain much higher than **death rates** so **the rate of natural increase** is high.

Birth rate is the number of live babies born per 1,000 of the population each year.

Death rate is the number of deaths per 1,000 of the population each year.

Natural increase is the difference between the birth rate and the death rate.

	Birth rate (per 1,000)	Death rate (per 1,000)	Natural increase
India	39	13	2.7%
UK	13	9	0.4%

Increasing population
- Large families
- Lack of contraception
- Slowing death rate as diseases are controlled
- Early marriage so long period of child-bearing

Declining population
- Smaller families
- Greater use of contraception
- Higher female literacy rates give women more opportunities
- Women marrying later and having careers

In richer countries populations are growing very slowly or even declining because falling death rates encourage people to have fewer children.

Factors affecting population growth

Agricultural change Children are needed to work on the land. As technology improves, machines take the place of manual labour so it is less important to have a large family.

Urbanisation People leave the countryside and move to towns. Children are more likely to go to school and families tend to be smaller in urban areas.

Education As levels of education rise, particularly among women, birth rates fall.

Women's status Countries where women are well educated and participate in paid work usually have a low birth rate.

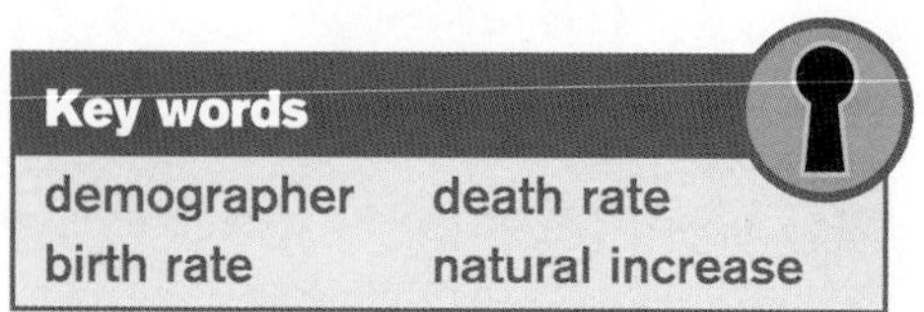

Key words
demographer
death rate
birth rate
natural increase

The demographic transition model

The demographic transition model (DTM) shows how the growth of population changes. Population growth is a balance between the number of live babies born and the number of people dying. The model is used to show how countries pass through different phases of population growth.

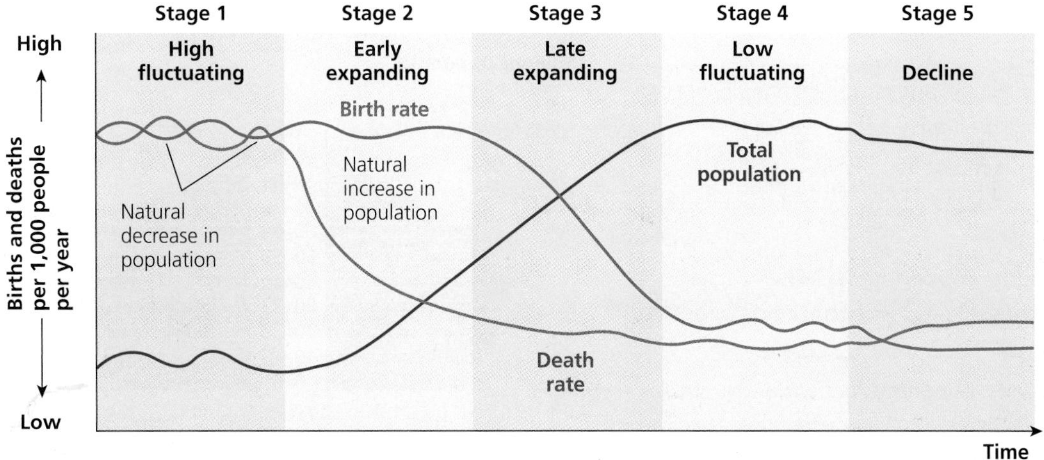

The demographic transition model

Stage 1 (high fluctuating) Shows a country or region before economic development. Both birth rates and death rates are high so there is a stable population size.

Stage 2 (early expanding) As the country starts to develop economically death rates fall because of improved medical care, clean water and better food supply. Birth rates remain high meaning that population size increases.

Stage 3 (late expanding) Death rates continue to fall but birth rates begin to fall as well because continued economic growth leads to improved education, better access to contraception and family planning advice. There may also be changes in social attitudes. Population size continues to grow but more slowly.

Stage 4 (low fluctuating) As living standards rise birth rates and death rates are both low, and in some cases there are fewer births than deaths. This reflects high levels of education and more women entering higher education. As more women work many choose to marry later and to have fewer children or even no children at all. This means population size remains stable.

Stage 5 (decline) This stage has recently been reached by some countries. Death rates are slightly higher than birth rates. Modern medicine is keeping people alive longer causing an ageing population. There are fewer people in the reproductive age range keeping birth rates low.

Remember this is a generalised model of population change. Not all countries will pass through all the stages in the same way or at the same speed.

Population structure

Population structure is the composition of a country's population by age and sex. It is usually shown as a **population pyramid** – which is like two bar graphs back to back, one for males and one for females. Age is shown in horizontal bars. The shape of population pyramids for different countries varies.

Life expectancy: average age people can expect to live.

Infant mortality: number of babies who die under the age of five years per 1,000 people.

Young dependants: children who are dependent on older **economically active** people.

Elderly dependants: people who are dependent on younger economically active people.

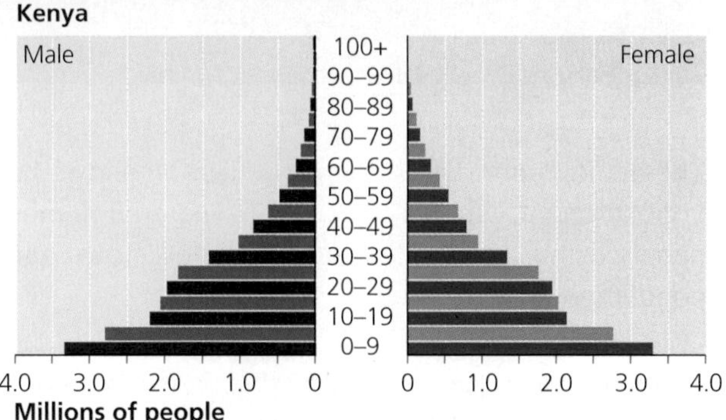

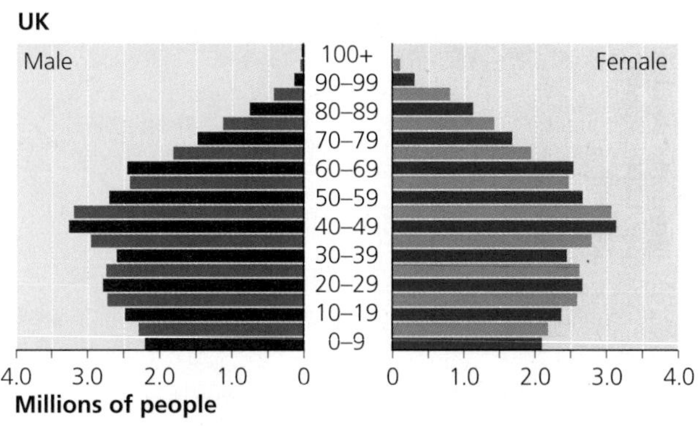

Population pyramids for Kenya and UK, 2009

The pyramid for Kenya shows many of the characteristics of poorer countries. The top of the pyramid is very narrow, showing there are few people over 65. There is a short life expectancy and few elderly dependants. The wide base shows a high birth rate and large numbers of children (young dependants). Because infant mortality and death rates are high, the pyramid narrows in successive age groups. The high birth rate and rapidly growing population is a problem for many poorer countries such as Kenya.

The pyramid for the UK shows the characteristics of a richer country. The narrower base shows low and falling birth rates so there are relatively few young dependants. The pyramid has a wide central part showing a large working population but also a wide top showing many people over 65, and more older women than men. The high number of elderly dependants is a growing problem for richer countries such as the UK.

Population pyramids are used to help predict changes in the population and plan for the future. They can be used to predict the proportion of elderly people in the population who will need health care, or the number of young people who will be economically active in the future.

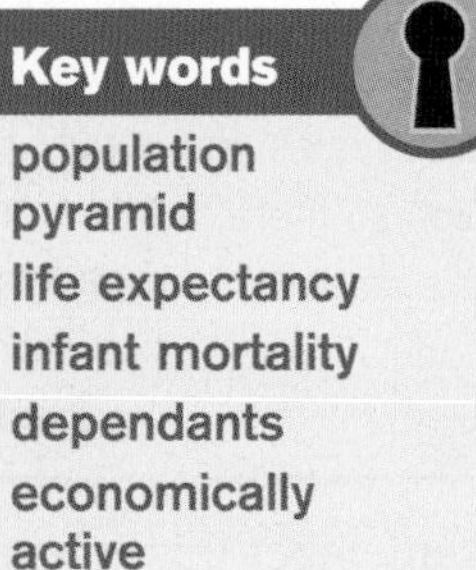

Key words
population pyramid
life expectancy
infant mortality
dependants
economically active

Problems of population growth

An area has a **sustainable population** if there are enough resources for people to continue to live there and maintain their standard of living without damaging the environment. Countries with low birth rates and low death rates and a stable or growing economy are the most sustainable.

If there are too many people living in an area for the resources available then the area is overpopulated. **Overpopulation** is not sustainable so it is important to reduce population in order to increase the resources per person and improve standards of living.

In some low-income countries population is increasing rapidly because birth rates are higher than death rates. An annual natural increase of about 2% may not sound much but is enough to double the population in 30 years.

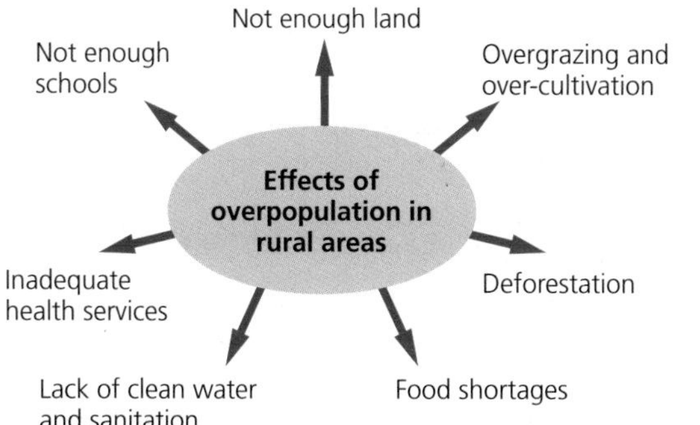

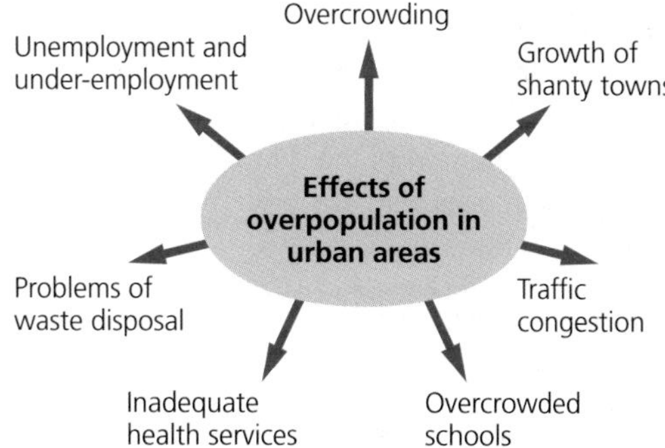

Population control policies

Many governments recognise the need to control their population in order to have a sustainable future. Mostly population policies aim to reduce birth rates by using persuasion and incentives as in India. However in China there are strict laws about family size.

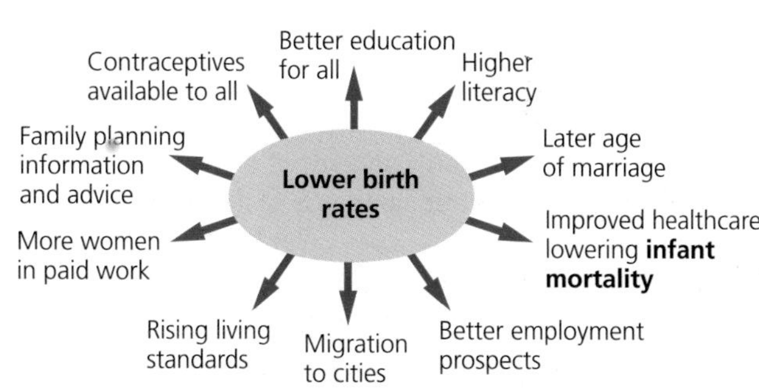

It is generally believed that economic development encourages people to have smaller families. This is because as a country becomes more prosperous there will be better health care and education as well as family planning advice, and access to **contraception**.

Key words

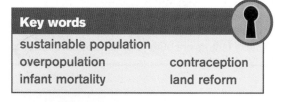

sustainable population
overpopulation
infant mortality
contraception
land reform

Case study: *non-birth-control policies*

Kerala, India

India has the second largest population in the world with over 1.1 billion people. Its population is growing quite rapidly and it is estimated that by 2050 India will have more people than China.

India has been trying to control its population for over 50 years. Despite some success it is still facing problems of overpopulation. If the population is not controlled there will be food and water shortages and more people living in poverty.

However, the state of Kerala in southern India has low birth rates similar to countries in Europe and the population is growing much more slowly than any other Indian state. Kerala is typical of a country in stage 4 of the demographic transition model with low birth rates and low death rates.

PAKISTAN
CHINA
BHUTAN
NEPAL
Delhi
INDIA
Kolkata
Mumbai
Arabian Sea
BANGLADESH
Bay of Bengal
KERALA
N
0 km 500

Kerala has achieved this because it has invested in social welfare schemes, particularly health, education and **land reform**. These have been put in place over the last 50 years by its strong socialist/communist government.

- **Education** – all children go to school until they are 16. Adult literacy classes are available in towns and villages. The literacy rate is 97% and more women go to university than men.
- **Health care standards** are excellent and life expectancy is 73 years. Improved access to doctors, vaccination programmes and better medical advice have reduced infant mortality.
- **Family planning policies** include providing free contraception and advice and educating people to understand the benefits of small families.
- **Higher incomes**, partly as a result of the redistribution of land (land reform), means there is less poverty and better standards of nutrition. This has also contributed to improved standards of health.

These changes in Kerala have led to:

- more women being literate, which means they are more likely to have paid work and to marry later
- couples now understanding about family planning so they can choose the number and spacing between children
- better health care and nutrition meaning children are less likely to die in infancy
- these factors encourage parents to have fewer children

Case study: *China's policy*

China's one child policy

The People's Republic of China has the world's largest population at 1,331 billion in 2009, about 20% of world population.

China is the only country to impose rules about family size as a way of controlling population growth.

In 1979, the government introduced the 'one child' policy which stipulates that couples may only have one child. This was encouraged by preferential access to housing, schools, and health services and punished with fines on each additional child and sanctions that ranged from forced sterilisation and pressure to abort pregnancy to discrimination at work.

Today the policy is still in place but has been modified so that couples from an ethnic minority or if both are only children themselves may have two children. In rural areas, a couple may also have a second child.

Critics say:

The one child policy has led to forced abortions and forced sterilisation

The same population reduction could have been achieved through voluntary means by encouraging couples to marry later and to space their children out, as was done between 1970 and 1979. Economic prosperity would also have resulted in lower birth rates

There is evidence of female infanticide (killing girl babies) because of the traditional preference for boys. Men now outnumber women in China by more than 60 million

Many Chinese people are unhappy as they want a larger family

Many only children are the focus of attention in their family and have become very spoilt; this is called the 'little emperor effect'

The one child policy has been successful because:

- population growth rates have fallen and there have been about 250 million fewer births than there would have been. It has also helped China's recent rapid economic growth
- problems resulting from overpopulation have been reduced, for example housing, health care, education and law enforcement

The future

- In 2008, a survey showed that over 76% of the Chinese population supports the policy and the government said the policy would remain in place indefinitely.
- The one child policy has slowed down population growth but there are still over 6 million more births than deaths every year in China, which means the population is still increasing and is likely to continue to do so until at least 2030.
- As single children grow up they have to support two parents and four grandparents; this is called the 4-2-1 problem. This is why if an only child marries they are allowed to have two children of their own.
- Young men may be unable to find wives as there are many more men than women.

Ageing populations

In many countries people are living longer and the proportion of older people is increasing, this is sometimes called the '**greying of the population**'. This is a problem because it means there are more older people who are dependent on a smaller **working population**.

In the EU, the proportion of people over 65 will nearly double by 2060, from 17% to 30%. There are now four people of working age for everyone over 65. By 2060 there will be only two.

In most EU countries the proportion of working people will decrease in the next 50 years. But the working population in the UK is likely to rise, because immigrants lift the fertility rate.

Reasons for ageing populations

- Increasing **life expectancy** – average life expectancy in the UK is 79 years.
- Low birth rates – in some countries birth rates are lower than death rates.

The consequences of ageing populations

- Growing market for leisure industries as older people spend on holidays and days out.
- House prices in popular retirement areas may rise.
- Cost of supporting older people through state pensions increases.
- Greater demand on medical services and long-term nursing care.

Solutions are difficult to find but some ideas are:

- Raising the age of retirement so pensions are paid later.
- Raising taxes on the working population to pay for care of the elderly.
- Providing incentives to encourage people to have more children.
- Encouraging immigration of young skilled adults into the workforce.

Most of these policies are unpopular.

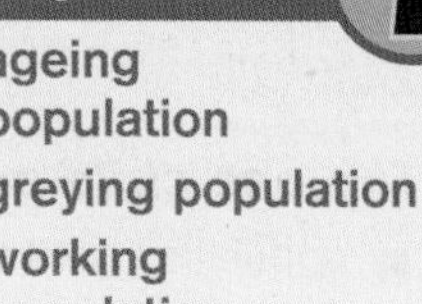

Key words

ageing population
greying population
working population
life expectancy

***Case study:** an ageing population*

Ageing population in France

France encourages families to have more children. It is a large country able to support a larger population and it needs younger people to fill gaps in the workforce.

- Parents are given three years' paid leave from their jobs (either mothers or fathers).
- Day care is subsidised for children under three.
- Full-time nursery care and schooling from age three is paid for by the state.
- Mothers who have three or more children are allowed to retire early with a pension.
- Families with three children or more get monthly allowances and tax concessions.

Birth rates are 1.9 children per adult which is higher that the European average but a third of the population will be over 60 by 2050.

Population movements

Migration

Migration is the movement of people from one region or country to another.

Emigration is people (emigrants) leaving a region or country.

Immigration is people (immigrants) entering a region or country.

Why do people move?

Push factors (cause people to leave)
- low wages so low standard of living
- lack of job opportunities
- poor quality of life
- lack of amenities, e.g. hospitals, schools
- conflict, e.g. civil war, oppression
- natural hazards, e.g. volcano, drought

Pull factors (attract people to an area)
- high wages and improved standard of living
- improved job opportunities, promotion
- better amenities and services
- improved quality of life
- better environment, no natural hazards
- freedom from oppression

There is usually more than one factor involved in the decision to migrate and it is usually a combination of both push and pull factors. Potential migrants also face a number of intervening obstacles that might prevent them moving, such as needing a visa to enter another country, not being able to speak the language, or not having enough money to travel.

Key words

migration
emigration
immigration
push factor
pull factor

Economic migration within the EU

There is freedom of movement for workers within the European Union (EU). UK citizens can move to other EU states to work and people from other EU countries can work in the UK.

Who migrates?

- Mostly young and single people, many of whom do not intend to stay permanently.
- Skilled, well-educated people who cannot find jobs in their own country.
- Some professional people, including doctors and dentists.

Benefits to the receiving countries

- Helps employers find workers when local people are not available or not willing to work, e.g. care workers, farm workers, builders and bus drivers.
- Local economies boosted because migrants spend money in local shops and services. Local councils can collect more council tax.
- Most migrants pay tax which contributes to services such as health, education and transport. The majority of migrants are young and single so make little use of these services.

Problems

- Local people may resent large numbers of newcomers.
- Pressure on rental properties and other services in the short term.
- Migrant workers often take low-paid jobs, so unskilled local workers lose out.

The future

In 2004, ten new countries joined the EU, including Poland and the Czech Republic. In 2007 Bulgaria and Romania joined the EU.

Many of the original 15 EU states have received a large number of migrants from countries that have joined the EU more recently. It is estimated that nearly one million workers have entered the UK and about half a million are still here. However, the economic downturn means that the UK is no longer such an attractive place for potential migrants and there are now fewer new arrivals and they are probably counter-balanced by those returning home.

Some transnational companies (TNCs) have closed factories in established EU countries, including the UK, and concentrated production in countries that have recently joined the EU. This means migrants may choose other EU countries in the future.

Refugees moving to the EU

Every year millions of migrants from outside Europe arrive in EU countries.

Some people leave their own countries because of war or violence, or because they are being persecuted as a result of their race, religion, nationality or political opinion. They arrive in Europe as **asylum seekers** and ask permission to stay because they are unable to return home safely. If this is granted they will officially become **refugees** and will be allowed to live and work in their new country.

Some people are looking for work and a better life and are **economic migrants**. EU countries have strict controls on migrants and if people enter, or stay, without permission they are **illegal immigrants**.

EU countries are trying to agree ways of dealing with immigrants because they all face problems. They need to work together to make sure only genuine refugees and migrants are able to live and work in the EU. At present different countries make their own decisions.

Impacts of refugees on host countries

- Refugees need housing when they arrive. They also need health care, they may not be able to speak the language well enough to find work and their children will need to go to school. All this must be funded by the government. In the long term, refugees and their children will be able to work and contribute fully to society.
- Some refugees are treated badly by people in their adopted country who do not understand what it is like to have to flee from your home. If a lot of refugees are housed in the same area it can cause conflicts between the newcomers and the local residents.
- The asylum process is expensive and can take months or years. Asylum seekers are usually held in detention centres and not allowed to work. Those who are unsuccessful will be deported.

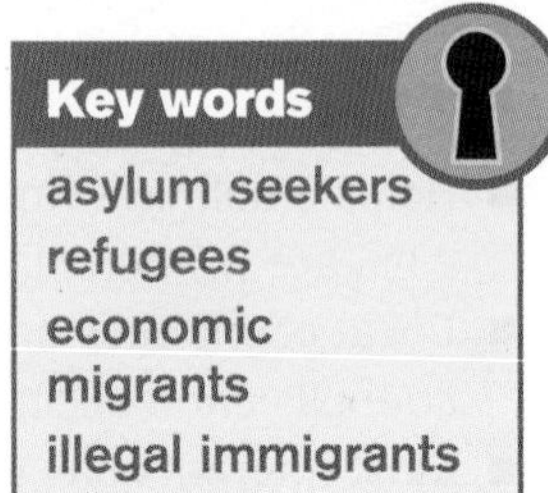

Key words

asylum seekers
refugees
economic migrants
illegal immigrants

Test yourself

1 Cross out the incorrect word(s) from those in italic in the following sentences.

(a) Migrants looking for work and a better life are called *asylum seekers/economic migrants*.

(b) Countries with a large proportion of older people are said to have *a youthful population/an ageing population*.

(c) Overpopulated areas have *too few/too many people* for the resources available.

(d) Birth rate is the number of live babies born *per 100/1,000/10,000* of the population each *month/year*.

Exam tip

Command words like 'describe' and 'explain' are very important (explain is the same as asking for reasons). Notice that the final part of this question asks for a named country you have studied. You must start your answer by naming a country or you will score very few marks.

Examination question

Foundation tier:

(a) Name one country you have studied where the government has tried to influence natural population change.

(i) Describe the ways the government tried to influence population change.

(ii) Explain some of the problems caused by this strategy. *(8 marks)*

Higher tier:

(b) Name a country you have studied where strategies have been introduced to influence natural population change.

Describe the strategies which were used and assess the extent to which these have been successful. *(8 marks)*

World urbanisation

The total population of the world is over 6 billion and expected to reach 7 billion in 2011. The proportion is much greater in richer countries such as Japan, Australia, the USA and the UK. Urbanisation occurred in these countries during the nineteenth and twentieth centuries. In poorer countries there is usually a smaller percentage of the population living in urban areas, although more people live in towns in South America.

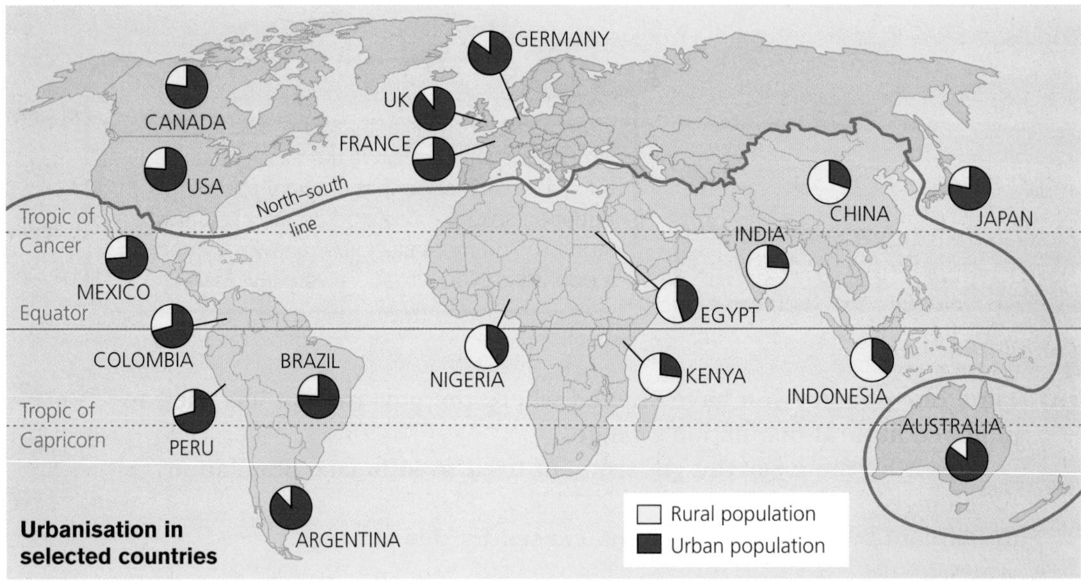

Urbanisation in selected countries

Urbanisation is the increase in the proportion of people who live in towns and cities. It is occurring on a global scale mostly in poorer countries.

Urban growth is the expansion of towns and cities so that they cover more land, as well as gaining larger populations.

Nineteen cities have over 10 million inhabitants and are known as **mega-cities**. Over 300 cities have over one million inhabitants and are known as **millionaire cities**.

Causes of urbanisation

Cities in poorer parts of the world are growing rapidly. This is due to two important processes:

1 **Rural–urban migration** The movement of people from the countryside to cities.
2 **Natural population increase** because birth rates are higher than death rates.

Rural to urban migration is the result of **push and pull factors**:

Pull factors attract people to cities because of the hope of work and earning money, better schools, healthcare, entertainment.

Push factors encourage people to leave villages because of poverty, few jobs except in farming, poor health facilities, few schools, little entertainment.

Land use in cities

In richer parts of the world cities tend to have distinct areas where similar types of **land use** are found.

Land use in a city falls into these categories:

Residential – Land used for housing.

Industrial – Land used for factories and other industrial buildings.

Open space – Land used for parks and playgrounds, and derelict or unused land.

Commercial – Land used for shops, offices, banks and other businesses.

Rural–urban fringe – Land on the outskirts where it changes slowly from a built-up urban area to countryside. It is the area where town and countryside merge.

The central business district (CBD) – The heart of the city, dominated by high-rise buildings occupied by shops, offices, banks and other **commercial functions**.

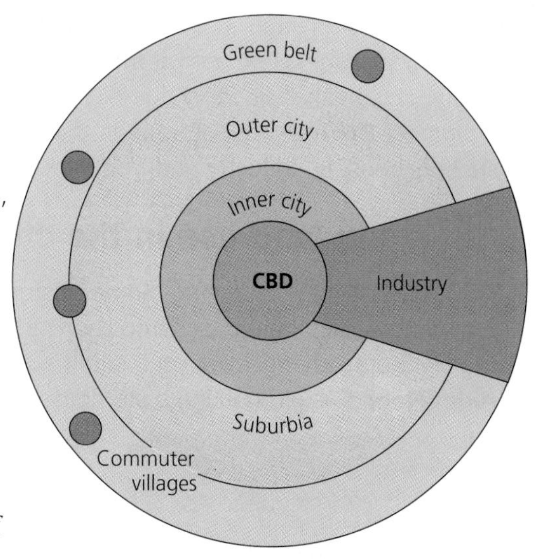

The model city – land use in a city in a rich country

The diagram shows a model city or the typical layout of land use in a city in richer parts of the world. This model is a useful tool, but remember that every town or city is unique. Towns and cities may show similarities to the model city but nowhere will be arranged just like the model.

The city skyline

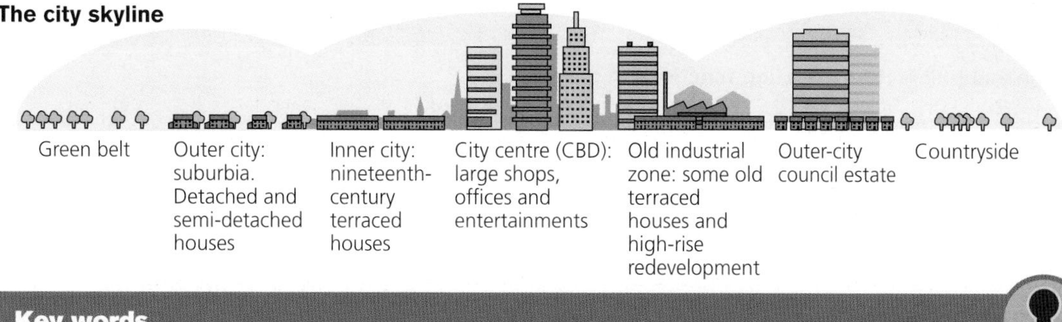

Key words

urbanisation	natural population increase	central business district (CBD)	clustering
urban growth	push/pull factors	commercial functions	quarter
mega-city	land use	leisure facilities	accessible
millionaire city			redevelopment
rural to urban migration			congestion

Revitalising the CBD

The central business district (CBD) is at the heart of the city. It is dominated by high-rise buildings occupied by shops, offices, banks and other commercial functions. There are often **leisure facilities** such as theatres, cinemas, night clubs, restaurants and pubs, and these may be **clustered** together in a **quarter**. All these functions group together in the CBD because it is the most **accessible** part of the city. People from all over the city and beyond can reach it easily. This pushes up the value of the land and also explains why there are so many high-rise buildings. The high cost of rents in the CBD means that some land uses are not found here, such as housing, industry and large areas of open space.

Changing land use in the city centre

The CBD is always changing. Some businesses move away as others move in. City councils are keen to keep businesses and shops in the city centre as this provides jobs and generates money. Some changes are on a small scale but in some city centres large areas have been **redeveloped** as old buildings are demolished to make way for new, modern developments.

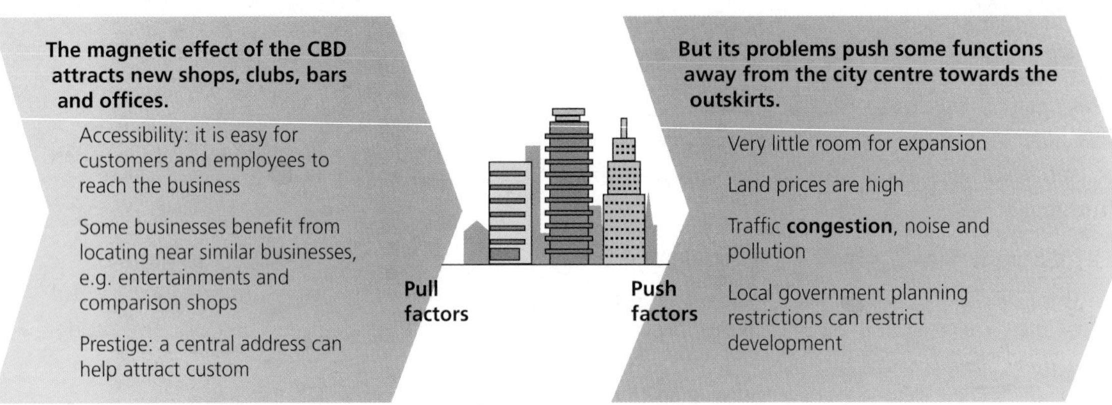

The magnetic effect of the CBD attracts new shops, clubs, bars and offices.
Accessibility: it is easy for customers and employees to reach the business
Some businesses benefit from locating near similar businesses, e.g. entertainments and comparison shops
Prestige: a central address can help attract custom
Pull factors
Push factors
But its problems push some functions away from the city centre towards the outskirts.
Very little room for expansion
Land prices are high
Traffic congestion, noise and pollution
Local government planning restrictions can restrict development

Push and pull factors affecting functions in the CBD

In many towns and cities the council has invested in schemes to improve the physical environment and make the CBD more attractive to people who live and work in the city, for visitors, and to encourage investment. Flats and apartments have been constructed either by converting old buildings or through purpose-built developments and cities are advertising 'city living' as a way to enjoy the convenience and facilities of central urban areas.

Changes include:

- new buildings for offices or apartments
- shopping malls
- public open spaces
- conversion of old buildings for a new use
- pedestrianised areas and one-way streets
- new types of transport such as trams (e.g. Manchester and Sheffield)

The inner city

Most inner-city areas lie close to the CBD and to industrial parts of the city. The **inner city** is an area of poor-quality, often nineteenth-century terraced housing. In the 1960s some of these old terraces were cleared as part of **urban redevelopment** schemes and replaced with council estates including high-rise flats. Today many inner-city estates are run down even though they are only about 50 years old.

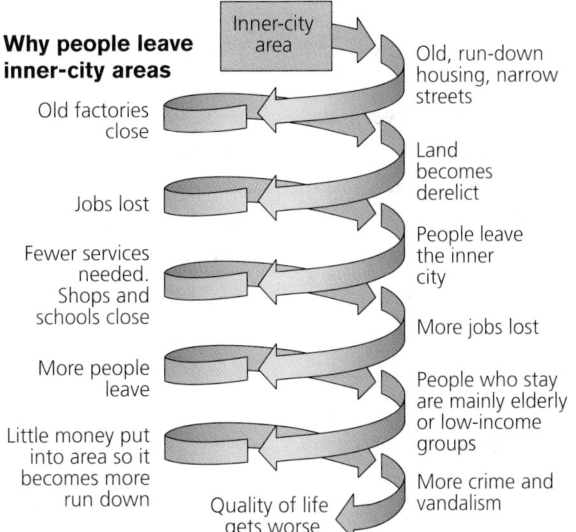
Why people leave inner-city areas
Inner-city area
Old, run-down housing, narrow streets
Old factories close
Land becomes derelict
Jobs lost
People leave the inner city
Fewer services needed. Shops and schools close
More jobs lost
More people leave
People who stay are mainly elderly or low-income groups
Little money put into area so it becomes more run down
More crime and vandalism
Quality of life gets worse

Cultural mix

Some inner-city areas have a significant number of immigrants who live where housing is cheap and where they are close to others with similar ethnicity or culture. This can result in **ethnic segregation**. Sometimes there are specialist facilities such as shops selling particular foodstuffs or religious buildings such as a mosque.

Government strategies to help the inner city

Successive governments have introduced strategies to revitalise inner-city areas and improve the **quality of life** for people who live there. Changes must improve the area for the long term; in other words they should be sustainable.

City Challenge This was a government strategy that ran from 1992 to 1998. There were 31 City Challenge Partnerships where local councils and private investors worked together to **regenerate** run-down areas. City Challenge projects aimed to improve housing, transport and the environment and to support businesses and create jobs; there was an emphasis on community projects, and training and education schemes. Examples of City Challenge schemes are: Hulme in Manchester and Deptford in London.

New Deal for Communities (NDC) The Labour government (elected in 1997) introduced a redevelopment strategy called the New Deal for Communities (NDC). Thirty-nine local partnerships were established in towns and cities throughout England. The emphasis was on local communities helping their own area. NDC focused on small areas and local community groups were encouraged to become involved with regeneration. This strategy is effective in supporting the multi-cultural nature of urban areas. Funding for NDCs continues until 2011.

Key words

inner city
urban redevelopment
ethnic segregation
quality of life
City Challenge
regeneration
New Deal for Communities

Demand for housing

Demand for housing in the UK is rising and towns and cities are now facing a housing crisis as there are not enough houses available.

More housing is needed because:

- Population is increasing and is expected to rise in the future. In 2000, the UK population was 58 million, by 2010 it had risen to 62 million and is expected to be over 75 million by 2050.
- The number of households is increasing and there are now over 26,000 households in the UK. One reason is increasing divorce rates.
- More people live alone. Young people may leave home before they get married and there are more older people living alone as life expectancy increases. About one third of single person households are aged over 65.

In 2007, the Labour government announced a target of building an extra three million homes in England by 2020. They wanted 60% to be built on **brownfield sites**. Brownfield sites are often in the inner city but gardens also count as brownfield sites and many house owners in the suburbs have sold off parts of their gardens for building land.

The greatest demand for new houses is in London and southeast England and this has led to pressure to build on **greenfield sites** on the outskirts of urban areas. There has also been pressure to build on **Green Belt** land so local councils can meet their house-building targets.

The coalition government elected in May 2010 scrapped house-building targets and said that Green Belt land should be protected and not built on.

	Brownfield sites	**Greenfield sites**
Advantages	• Easier and quicker to get planning permission • Services like electricity, gas, water and sewerage already in place • Roads and other transport links already in place • Closer to the city centre to work or for shops and entertainment	• Site has not been built on before so it is quicker and easier to prepare • Land is cheaper on the outskirts • Surrounding area may be attractive making it easier to sell larger and more expensive houses which make more profit for the developer
Disadvantages	• Land may be expensive to buy or rent if it is near the city centre • Site may be polluted and expensive to clean up • Site may not be very large, perhaps only big enough for a few houses • Surrounding area may not be attractive making it difficult for builders to sell houses	• Countryside is lost to houses and roads • Wildlife is threatened

Key words

brownfield site
greenfield site
Green Belt

Traffic in cities

Traffic congestion is a major problem in towns and cities all over the world. Roads are congested by more and more cars (75% of all households in the UK now own at least one car) and by thousands of delivery vehicles and public transport, particularly buses.

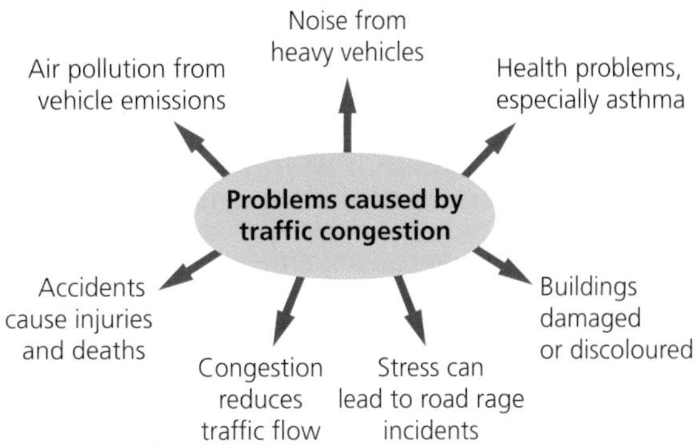

Problems caused by traffic congestion
- Noise from heavy vehicles
- Health problems, especially asthma
- Buildings damaged or discoloured
- Stress can lead to road rage incidents
- Congestion reduces traffic flow
- Accidents cause injuries and deaths
- Air pollution from vehicle emissions

Tackling traffic problems

- Create one-way systems and pedestrian only areas.
- Designate cycling lanes to encourage people to leave their car at home.
- Build underground or multi-storey car parks.
- Build new roads to take traffic around the city centre (ring roads) or even over the centre on urban motorways and flyovers.

Integrated public transport schemes

Manchester and Sheffield are two cities that have successful, modern tram systems connecting the suburbs with the city centre and industrial areas and linking with other types of public transport such as buses and trains. Large car parks on the outskirts encourage people to park and take public transport into the centre.

Park and Ride

In York, Cambridge, Oxford and many urban areas, Park and Ride schemes encourage visitors to park their cars on the outskirts and take a bus into the centre rather than driving into congested city centres.

Case study: *congestion charges*

London Congestion Zone

London introduced congestion charges in 2003 to try to reduce traffic congestion in central London. The scheme was extended into parts of west London from 2007 to 2010.

Motorists are charged £10 per day to drive into the zone between 7am and 6pm. The system is controlled using Automatic Number Plate Recognition and drivers are fined between £60 and £180 for non-payment of the congestion charge.

Studies show 70,000 fewer cars enter the original congestion zone each day compared to pre-charging levels, and 30,000 fewer cars entered the western extension. However, even with this reduction in vehicles the area remains congested so the scheme is only partly successful. Congestion charges generate about £140 million a year, which is invested in improving other aspects of London's transport.

Cities in poorer countries

Cities in poorer parts of the world are growing fast. Industrial growth usually occurs in cities and provides employment. Cities provide better services than most rural areas, particularly health care and education.

However, rapidly growing cities also have many problems, particularly in countries where there may be little money available to invest in city planning.

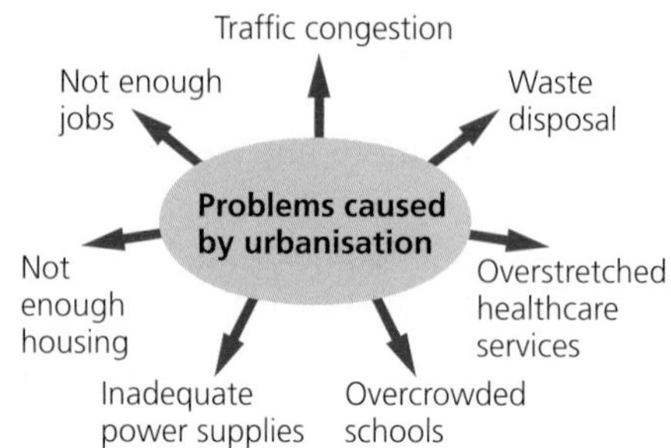

Squatter settlements

Rural–urban migration in poorer countries usually leads to the development of **squatter settlements** on the outskirts of the city or on land that is not suitable for housing. This type of settlement can cover a large area and may house over 30% of the population.

The poorest urban inhabitants, often recent migrants, live in squatter settlements that have different names in different countries, e.g. bustee (India), favela (Brazil), barriada (Peru).

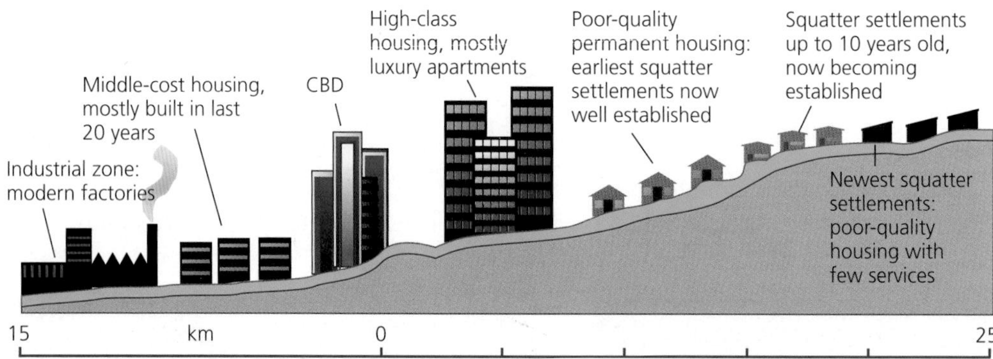

Cross-section through a city in a poorer country

The buildings are usually made with locally available materials, such as mud bricks or wooden walls and plastic or corrugated-iron roofs.

Living conditions

- Very crowded areas with high population densities.
- Few basic services such as clean water and sanitation, waste disposal, electricity supplies or paved roads.
- Inadequate health care, high infant mortality and a short life expectancy.
- There are not enough 'formal' jobs and squatter areas may be a long way from the CBD, so many people earn what they can from informal occupations such as street trading.
- The location is often dangerous or unpleasant.

Improving squatter settlements

Local authority schemes for new housing

City authorities rarely have enough resources to tackle the problems of large squatter settlements. In some cities, such as Lagos in Nigeria, the authorities have built apartment blocks to re-house people. In many cities this is not an affordable option.

Site and service schemes

In some countries the authorities have tried to help migrants to the city to build their houses according to careful guidelines. This can be successful on a small scale.

An area of land is found that is not too far from work places in the city. It is divided into individual plots of land by the authorities. Roads, water and sanitation may be provided. Newcomers can rent a plot of land and build their own house, following certain guidelines. As time goes by and they have more money they improve their house.

This approach was used in Lima in Peru. New townships such as Villa el Salvador were laid out in the desert surrounding the city.

Advantages

- cheaper than building new flats
- houses are better built because of guidelines
- water and sanitation supplies reduce risk of disease

Disadvantages

- difficult to provide enough plots of land for the huge numbers needing housing
- families without employment cannot afford to pay rent

Self-help schemes

Once people have built a house, they are likely to improve it when they can. They will only do this if they are confident that they will not be thrown off the land, so they need to have legal ownership of the land.

Self-help schemes are important in almost all big cities in poorer countries. People improve their houses slowly, e.g. replacing mud walls with bricks or breeze blocks. The house may gradually be enlarged, building more rooms and then adding upper floors.

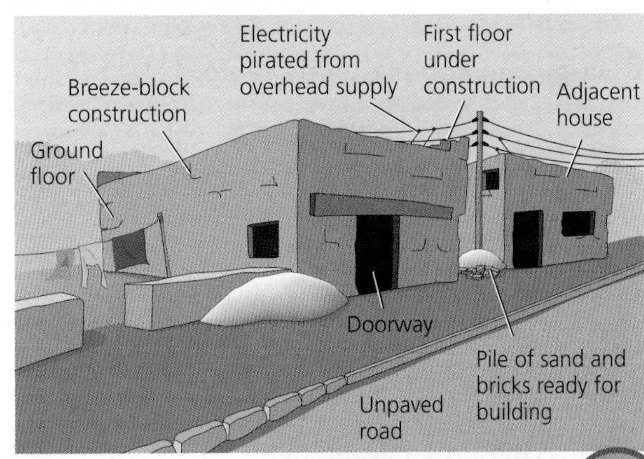

Self-help housing scheme

Key words

squatter settlement
site and service scheme
self-help scheme

Later, city authorities may provide clean water from standpipes, and help with sanitation and waste collection. Commercial bus operators will start bus services, and health centres may be built by the local community. People work together to improve their area and over time it changes from a poor, illegal settlement to a legal, medium-quality housing area.

Non-governmental organisations (**NGOs**) such as Oxfam and Save the Children encourage people to help themselves through **micro-credit schemes**. They provide small grants or loans to the poorest people along with training and advice to help people start their own business.

Key words

NGOs (non-governmental organisations)
micro-credit schemes

Case study: *squatter settlement redevelopment*

Rocinha, Rio de Janeiro

Rocinha is an example of a squatter settlement that has been gradually improved and redeveloped largely by the people who live there. It is the largest favela in the city with an estimated population of 250,000.

- It is built on a steep hillside.
- The original buildings were flimsy structures made from materials such as wood and corrugated iron.
- Today almost all the houses are made from concrete and brick. Some buildings are three and four storeys tall and all have sanitation, plumbing, and electricity.
- The infrastructure in Rocinha has developed so there are roads and water supplies.
- Health centres and schools have been built.
- Hundreds of local businesses have been established, including shops, bus services and even a locally based channel, TV ROC. Many people work in central Rio and others have jobs in the informal sector, including selling food and souvenirs on the beach.

Test yourself

1 **Copy and complete this table.**

	Causes	Consequences
Urbanisation	(1) (2)	Overcrowded housing
Counter-urbanisation	(1) (2)	Expansion of rural settlements

2 **Describe how people living in squatter settlements can try to improve their own lives.**

3 **What are the problems caused by traffic congestion in towns and how can they be solved?**

4 **In what ways can planners improve central areas of cities?**

Environmental problems caused by rapid urbanisation

Social and economic problems caused by rapid urbanisation are usually worst for those in squatter settlements. Environmental problems caused by increase in population and industrialisation are likely to be experienced throughout the city.

Environmental problems

Air pollution

- Traffic emissions, particularly carbon dioxide, cause severe air pollution.
- Factories emit pollutants such as sulphur dioxide and nitrogen oxide. This often causes a haze of pollutants in the air above the city.
- Power stations burning fossil fuels add carbon dioxide to the atmosphere.

Solid waste

- Domestic rubbish is often uncollected especially in squatter settlements. It is unsightly, smells and can cause health problems.
- Huge landfill sites develop on the edge of the city for the collected waste.
- Electronic waste (e-waste) is a big problem as it needs specialist disposal methods. Sometimes e-waste from other parts of the world is taken to cities in poorer countries for disposal. E-waste may be broken up by people who try to recycle valuable parts such as gold and platinum. Unfortunately, they may also expose themselves to high levels of toxic chemicals such as cadmium and lead.

Water pollution

- Rivers used as open drains may carry untreated sewage.
- Industrial waste is dumped in rivers.
- Chemicals may be disposed of in rivers. This toxic waste can be dangerous.

Managing environmental problems

Governments in many poor countries are struggling to provide basic services for their growing populations, and tackling environmental problems may be seen as less important.

- Water pollution could be tackled by cities investing in improved water supply and sanitation systems but this is expensive.
- Reducing air pollution needs legislation to stop factories emitting high levels of polluting gases. Even where laws exist, there are often not enough people employed to enforce them.
- Waste disposal is one of the biggest problems. Poor people may earn their living by sorting through the rubbish on landfill sites and selling or recycling what they can.
- Recycling is important and people, often children, can earn a living by recycling goods. For example, they collect and sort glass, paper and plastic and sell it to companies that reuse it.

Sustainable cities

About half of the world's population now live in urban areas. Most cities are unsustainable, in other words they consume raw materials, energy and water and produce harmful waste. As cities grow they take up more land and damage the environment. The planet cannot support this and we now realise ways must be found to reduce the impact cities are having and make them more **sustainable**. Making urban areas sustainable means changing people's lifestyles and the way they think about the environment as well as changing the way cities are planned.

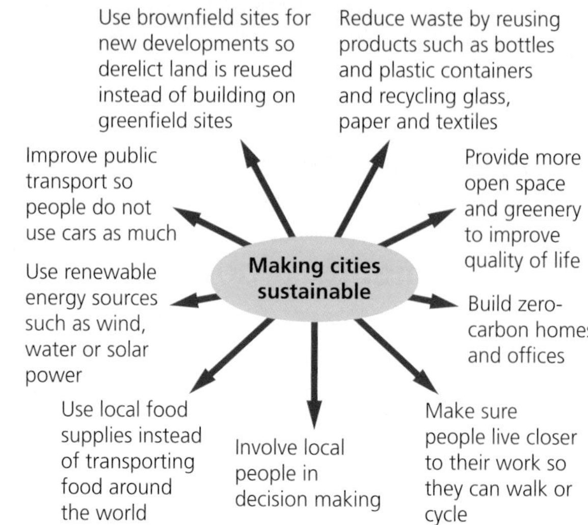

Making cities sustainable
Use brownfield sites for new developments so derelict land is reused instead of building on greenfield sites
Reduce waste by reusing products such as bottles and plastic containers and recycling glass, paper and textiles
Provide more open space and greenery to improve quality of life
Build zero-carbon homes and offices
Make sure people live closer to their work so they can walk or cycle
Involve local people in decision making
Use local food supplies instead of transporting food around the world
Use renewable energy sources such as wind, water or solar power
Improve public transport so people do not use cars as much

Zero-carbon living

About 25% of the UK's carbon emissions come from homes. It is possible to reduce this by improving insulation and installing double glazing to prevent heat loss and by encouraging people to reduce the amount of energy they use, e.g. by switching to low-energy light bulbs, turning off stand-by switches and turning down heating. **Carbon neutral** homes are those where carbon emissions do not add to the net amount of carbon dioxide in the atmosphere.

The Labour government wanted all new homes in the UK to be **zero-carbon** by 2016. This means not releasing any carbon dioxide at all into the atmosphere. This is a big challenge as most electricity is still generated by burning fossil fuels. New homes will need to make their own energy, e.g. by having solar panels and using boilers fuelled by biomass.

Eco-towns

In July 2009, the Labour government announced plans to build ten **eco-towns** in England. Each will have 4–5,000 homes. This will be reviewed by the coalition government.

All houses will be zero carbon and generate their own energy from renewable sources. Residents will be able to sell surplus energy to the national grid.

Towns will have smart meters to track energy use, community heat sources and charging points for electric cars.

There will be efficient public transport, cycle routes and footpaths, and shops and a primary school within easy walk of every home.

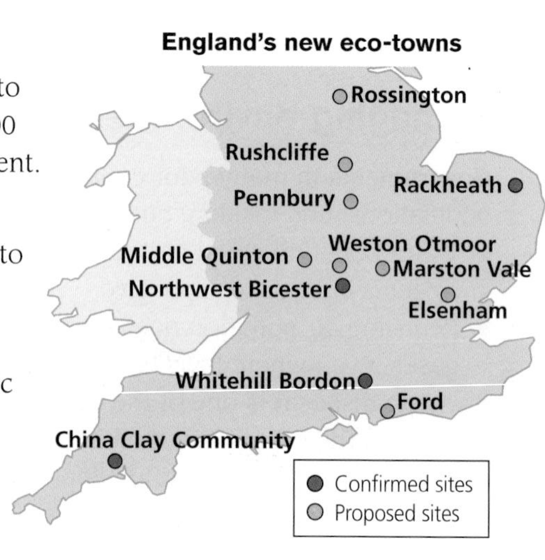

Some people think eco-towns are a good idea because:

- towns are planned to be sustainable
- they will provide thousands of new houses and many will be 'affordable' homes for lower income families
- they will help to tackle climate change by reducing emissions of greenhouse gases

Some people think eco-towns are a bad idea because:

- they will be built on greenfield sites using valuable countryside and green belt land
- too few houses will be built to meet demand
- they will generate more traffic, which will generate more carbon emissions

Case study: *sustainable urban living*

Bed-ZED, Wallington, South London

In 2002, the Beddington Zero Energy Development or Bed-ZED was opened. It is an experiment in zero carbon living. There are 100 households living in apartments that have been designed to save as much energy as possible.

- Wind turbines and biomass boilers generate power.
- Buildings are well insulated.
- Rainwater is collected and reused.
- Residents use public transport or bikes and also have shared cars.
- Residents try to buy locally produced food and recycle their waste.

This community has shown that it is possible to make significant savings on heating, electricity and water consumption and they have reduced their **ecological footprints**. However, they have still not achieved the aim of 'One Planet Living'. This is because of the environmental impact that occurs outside their homes, e.g. at their schools and workplaces, and because many of the goods and food they buy is not local.

Bed-ZED is in the London Borough of Sutton and in 2005, Sutton Council made a commitment to become a 'One Planet Living Borough' by 2025. They will be trying to reduce the ecological footprint of the whole borough and apply the lessons learnt from Bed-ZED.

Key words

- sustainable city
- carbon neutral
- zero-carbon homes
- eco-town
- ecological footprint

Bed-ZED housing

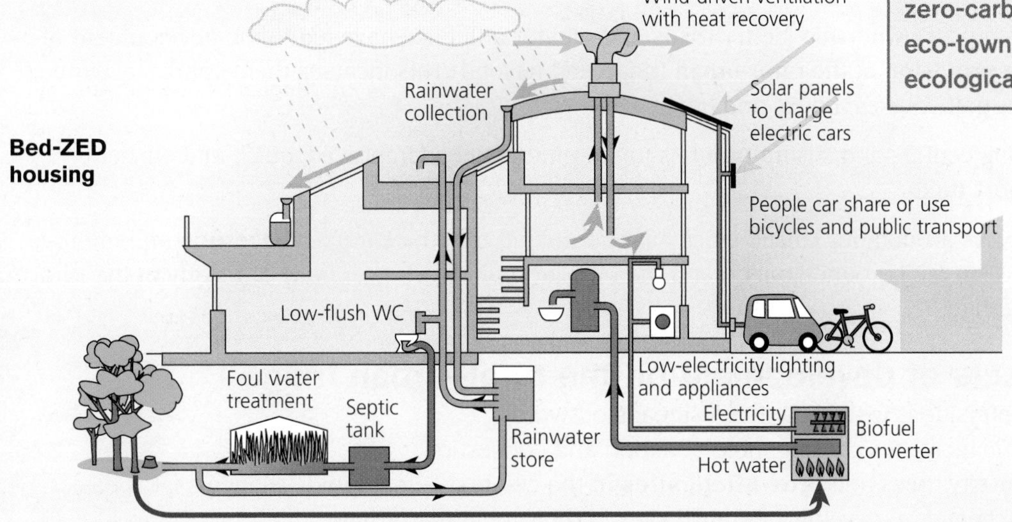

The rural–urban fringe

Land on the outskirts of the town changes slowly from a built-up urban area to countryside. This area where town and countryside merge is known as the **rural–urban fringe**. It is a transitional zone that has both urban and rural land uses.

The growth of towns and cities outwards is called **urban sprawl**. Land here is targeted for new building developments because:

- land on the outskirts is cheaper and more readily available
- main roads and motorways make this area accessible
- developers prefer **greenfield sites**

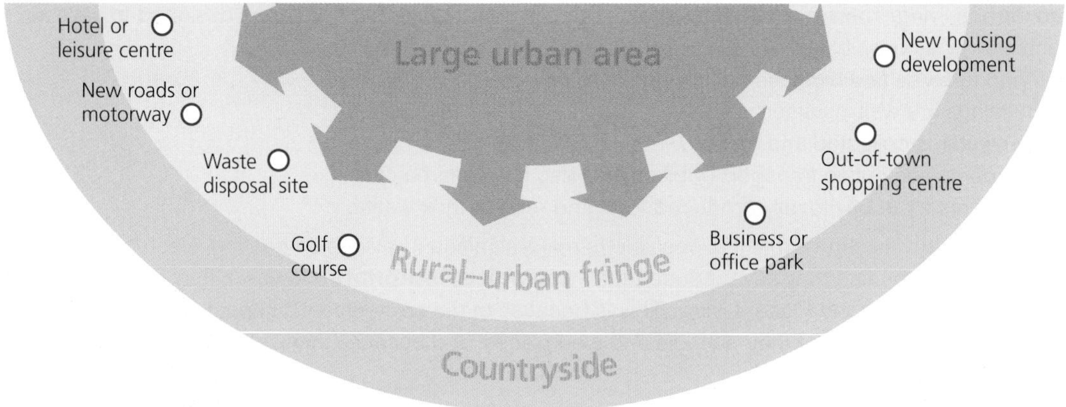

Threats to the rural–urban fringe

Many cities have large **out-of-town retail outlets**, sometimes called **retail parks**, such as the Trafford Centre in Manchester and Bluewater Park in Kent.

Increasing car ownership, better roads and greater affluence have led to the development of **leisure provision** in the rural–urban fringe and beyond. This includes theme parks, activity centres, golf courses, hotels and health centres.

Shopping centres and leisure facilities for commuters need improved, quick and efficient **transport links**.

Motorways around our largest cities have been built to reduce traffic congestion and improve access. The M25 around London and the M60 around Manchester both lie mostly in the rural–urban fringe.

Impacts of development on the rural–urban fringe

- Countryside is lost as the town spreads outwards.
- Traffic increases causing more pollution and congestion.
- Inner-city areas and **brownfield sites** in the city may remain undeveloped as it is easier and cheaper to develop greenfield sites in the rural–urban fringe.

Suburbanised villages

During the last 30 years in richer countries people and businesses have moved away from large urban centres to small towns or villages. This process is known as **counter-urbanisation**. People move for a number of reasons or **push factors** and are attracted to villages or small towns by **pull factors**.

Some villages have grown rapidly through **urban to rural migration**. New houses have been built and older ones renovated. In some places, the total population has more than doubled, bringing with it great changes to the village. These **suburbanised villages** are usually within easy reach of a large urban area because many people who choose to move are **commuters** who travel back to the city each day to work.

As well as commuters, people may choose to move out of the city when they retire. Changes in technology allow some people to work from home and they can choose to live where they want, which may be in a rural area.

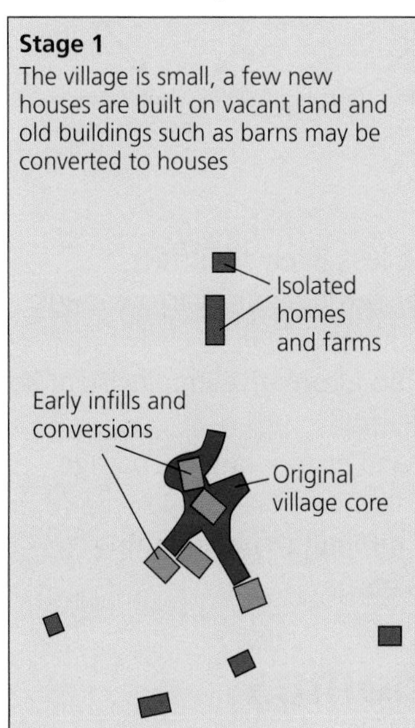

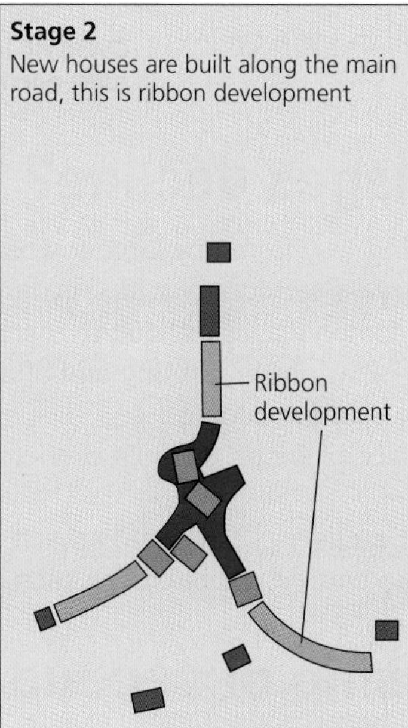

Stage 2
New houses are built along the main road, this is ribbon development
Ribbon development

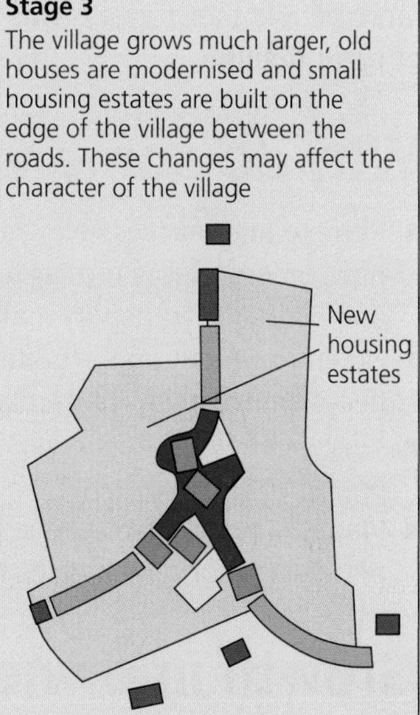

Stage 3
The village grows much larger, old houses are modernised and small housing estates are built on the edge of the village between the roads. These changes may affect the character of the village
New housing estates

A suburbanised village

Why do some villages expand?

- Accessible – close to main road or a railway station.
- Not too remote – within reach of a range of employment and services.
- Attractive – village green, church or old pub.
- Landscape – surrounded by attractive countryside.
- Few planning restrictions – possible to build new houses or renovate older buildings.
- Village services – shops, primary school, a pub, a doctor and other services.

Key words

rural–urban fringe
urban sprawl
greenfield site
out-of-town retail outlet
retail park
brownfield site
counter-urbanisation
push/pull factors
urban to rural migration
suburbanised village
commuter

Remote rural areas

Rural depopulation

Some rural areas are losing population because people are moving out and fewer people are moving in. There are often few job opportunities in rural areas so people move to towns to find work. Many of the people left behind are older. As numbers decline services such as shops and schools close down and this may mean even more people move away. This can cause a cycle of **rural decline**.

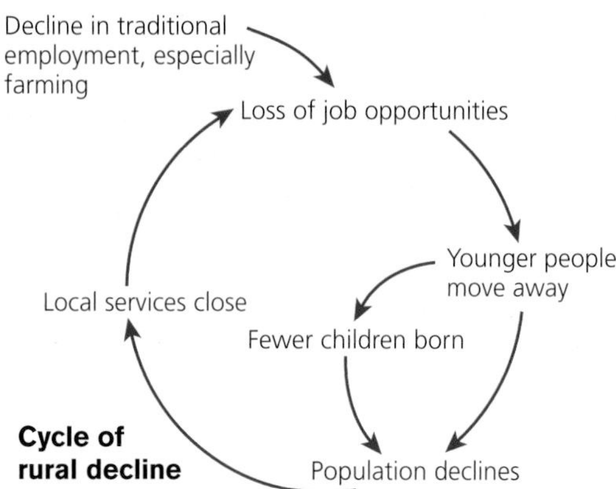

Why do some villages decline?

- Remote and inaccessible – a long way from any large towns so less likely to attract newcomers as it is too far to access services provided by larger settlements. Narrow roads and no public transport make some villages difficult to reach.
- No employment opportunities – few jobs in farming and often no alternative employment so people move away, particularly younger people.
- Very small with no services – lack of shops, schools, a doctor and even a pub encourage people to leave.
- Planning restrictions – in some areas, e.g. National Parks, it is difficult to get planning permission for new houses or to convert old buildings such as barns.

Growth in ownership of second homes

In attractive rural areas, such as National Parks, and on the coast houses may be bought as a **second home**, often by people with well-paid city jobs. In areas like Cornwall, the Lake District and Snowdonia, many houses have been bought by people who only use them at weekends or for holidays. Many houses may also be rented out during the summer. In some villages more than half the houses may be second homes. This causes problems because:

- house prices are pushed up and local people may not be able to afford to buy a house
- houses are left empty for long periods and the owners are not part of the village community
- owners may bring their own food with them and not use local services

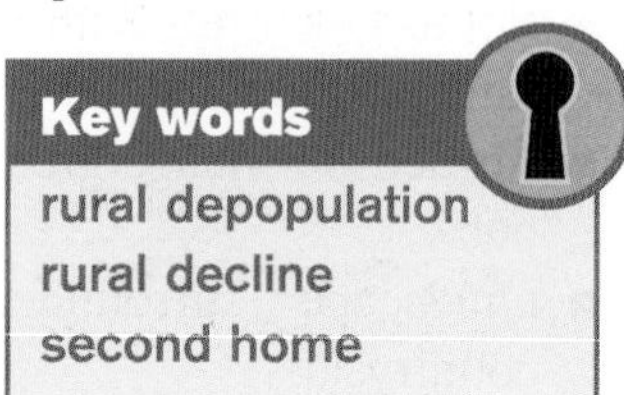

Case study: *a rural area in the UK*

Cornwall

Cornwall is well known as a holiday destination; there are over 300 beaches and the climate is warmer than in other parts of Britain. However, it is also experiencing rural depopulation and rural decline, particularly in inland areas which are remote and sometimes difficult to get to.

Rural depopulation in inland Cornwall

- Traditional industries have declined, particularly farming, fishing, tin mining and extraction of china clay. There are few alternative employment opportunities except tourism, which is often seasonal with low pay.
- It is remote and a long way from London – the nearest large city is Exeter which is over an hour's drive from Truro.
- There is an ageing population as younger people move out to find jobs elsewhere.
- Services such as shops, schools, petrol stations, banks and bus routes are closing as there are too few people to support them.

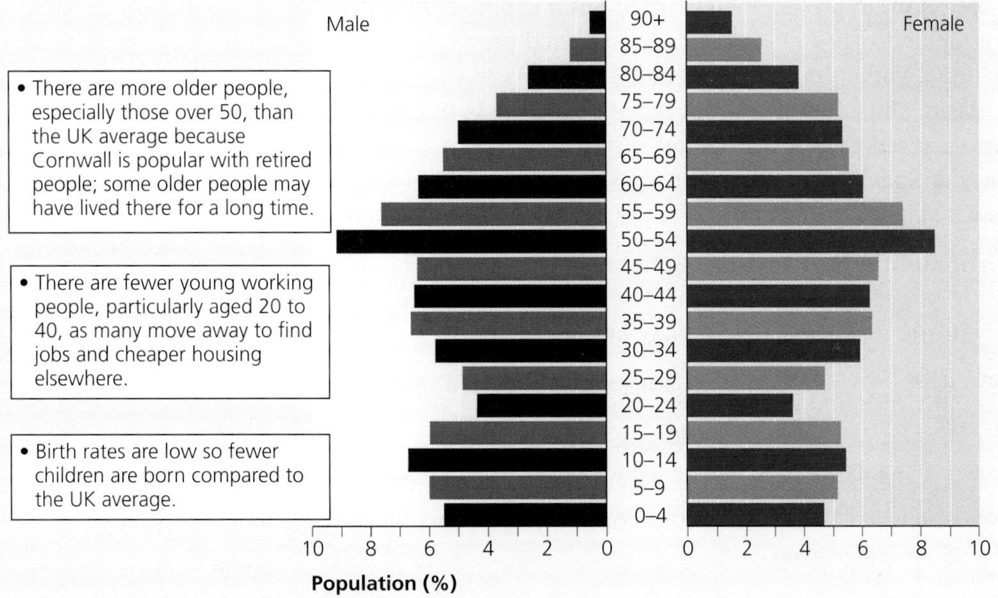
Male
Female
90+
85–89
80–84
75–79
70–74
65–69
60–64
55–59
50–54
45–49
40–44
35–39
30–34
25–29
20–24
15–19
10–14
5–9
0–4
• There are more older people, especially those over 50, than the UK average because Cornwall is popular with retired people; some older people may have lived there for a long time.
• There are fewer young working people, particularly aged 20 to 40, as many move away to find jobs and cheaper housing elsewhere.
• Birth rates are low so fewer children are born compared to the UK average.
10 8 6 4 2 0
0 2 4 6 8 10
Population (%)

Population pyramid for inland Cornwall

Second homes

Coastal areas of Cornwall have a lot of houses that have been bought as second homes. In some places, such as Port Isaac and Boscastle, it is thought that 60–80% of properties are second homes. The owners may only visit these houses a few times a year and may not even use local shops or services and they are not part of the local community. Cornwall has a high number of local people who are unable to buy their own homes because wealthy outsiders push property prices up so there is a lack of affordable housing.

Making rural living sustainable

Problems such as depopulation and loss of rural services need to be addressed. It is important that new developments are sustainable, in other words that changes that help the area today will not compromise the needs of future generations.

Natural England was set up by the government in 2006. This organisation is responsible for protecting our landscape and wildlife.

Conserving resources and protecting the environment

It is important to protect the environment and conserve our resources for the future. Many beautiful landscapes in England are protected as:

- National Parks: 10 in England and 3 in Wales
- Areas of Outstanding Natural Beauty (ANOB): 34 in England
- nature reserves: 224 in England
- Sites of Special Scientific Interest (SSSI): over 4,000 in England
- green belt: 14 in England

Environmental Stewardship Scheme – Farmers and landowners can apply for grants if they agree to protect the environment. The money is provided through the EU Common Agricultural Policy (CAP). (See page 109.) Grants are given for:

- repairing and restoring hedges
- planting trees and looking after woodlands
- maintaining hay meadows and grasslands
- growing plants that provide seeds for birds or nectar for insects

Farmers receive £30–60 per hectare for five years depending on the amount and type of stewardship undertaken.

Supporting the needs of the rural population

People in the countryside need affordable housing, jobs and services, including shops and public transport. In the past there were government schemes to help village shops, community projects and rural transport schemes but funding from central government is no longer available. House prices are often pushed up in rural areas by people buying second homes or moving into countryside areas to retire.

The Rural Development Programme – This is part of the EU Common Agricultural Policy (see page 109). Some CAP funding is earmarked for rural development to help to improve the quality of life, encourage diversification and increase job opportunities such as within the tourist industry.

Commercial farming in East Anglia

The low-lying land of East Anglia between The Wash and the Thames Estuary is suited to **arable farming**.

- The main crops are cereals, particularly wheat and barley, sugar beet and potatoes.
- Livestock farming is important in some areas, mainly pigs and chickens.

Farming in East Anglia is **commercial** as the crops are all sold for profit and **intensive** as there are high yields per hectare as a result of high inputs of technology, including machinery and fertilisers.

About 50,000 people work on farms but many more jobs depend on it, including workers in livestock feed manufacture, transport, and agricultural research and development.

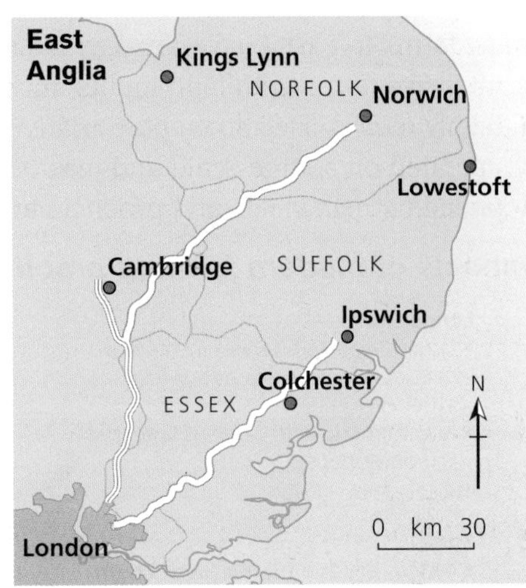

Physical factors favouring farming

- Flat, low-lying land means large machines such as combine harvesters can be used. Hedges have been removed to make huge fields suited to the large machinery.
- The long growing season from April to September and warm summer temperatures (average 18°C) help crops to grow and ripen producing high yields.
- Average annual rainfall is 600 mm rain which is enough for crops but not too much.
- The soil is deep and fertile (mainly boulder clay) which helps crops to grow.

Human factors favouring farming

- The area is close to large urban areas, including London, which provide a market for the crops.
- Some crops, particularly sugar beet, are processed locally. Sugar beet is easier to transport as raw sugar.
- Main roads and motorways allow crops to be transported quickly to markets in East Anglia and London.

Diversification

Diversification is when farmers develop business activities other than farming on their land. They do this to increase their income.

Key words
- arable farming
- commercial farming
- intensive farming
- diversification

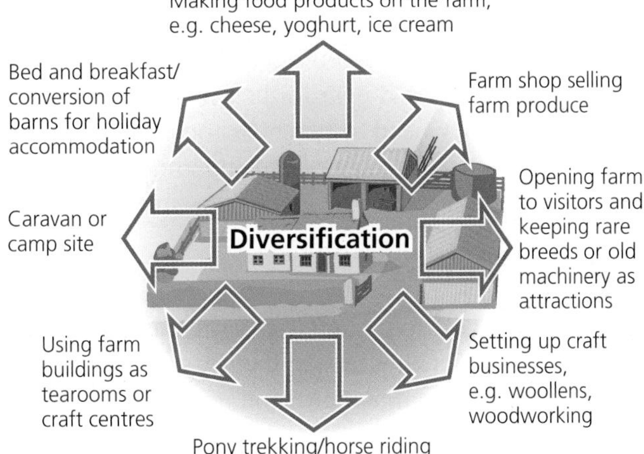

Agri-business

Many farms in East Anglia are agri-businesses, which are:

- intensive farms with high inputs, including chemical fertilisers and pesticides.
- highly mechanised so employ relatively few people.
- operated on a large scale and may be owned by a company and run by a farm manager.
- providing high yields and products to supply supermarkets or food processing companies.

Impacts of modern farming practices

Landscape
- larger fields with fences instead of hedges – more efficient but less attractive
- fewer crop varieties means landscape is monotonous

Water courses
- nitrates from fertilisers get into streams and lakes causing eutrophication
- phosphates from animal manure (slurry) pollute water supplies
- draining wetlands reduces wildlife and increases flood risk

Soil
- hedges act as a natural wind break and roots bind soil – when removed there is more soil erosion
- large machines are efficient but are heavy and can cause soil compaction

Wildlife
- chemical pesticides destroy insect pests so that yields are higher – but they also kill bees and butterflies needed for pollination and natural predators such as ladybirds
- removal of habitats such as hedges and ponds leads to loss of species

Farming and supermarkets

Many farmers sell their products to supermarkets but may not get a good price because:

- they pay farmers low prices to keep prices in their shops competitive
- farmers may be in competition with overseas farmers whose production costs are lower

Farmers try to reduce their own costs to increase their profits. The cheapest way to rear chickens is in cages and many consumers are not willing to pay extra to buy free-range eggs. Battery cages will be banned in the EU in 2012. Marks & Spencer and Waitrose now sell only free-range eggs.

Farming and food processing firms

Farmers' products may need to be processed before they can be sold. The food processing industry in East Anglia is worth £3 billion a year. Some products such as sugar and frozen vegetables are processed near to their raw materials. Some products are processed near to their market as they need to reach customers quickly, e.g. bread and pastries. Food processing companies are under pressure from supermarkets to keep their prices down so they buy raw materials from farmers as cheaply as possible.

Organic farming

Organic farmers aim to protect the Earth's resources and to produce safe and healthy food sustainably. They employ certain methods:

- crop rotation, animal manure, and no chemical fertilisers

- natural predators, such as ladybirds, which eat greenfly and are used instead of chemical pesticides
- weeds are controlled mechanically instead of with synthetic herbicides
- no genetically modified (GM) crops are grown

Advantages of organic farming

- Soil is not damaged, its fertility is protected and there is little or no soil erosion.
- Streams, lakes and rivers are not polluted.
- Food has more nutrients and tastes better.
- It protects habitats and encourages biodiversity.
- It reduces greenhouse gas emissions caused by the manufacture of chemical fertilisers.
- All animals are well cared for.

Disadvantages of organic farming

- Lower yields because no chemical fertilisers are used and crops are more likely to be eaten by insects or affected by disease.
- It is more expensive.
- Organic crops are grown in their correct season, and may not be available at other times.
- Organic cattle produce twice as much methane as cattle reared traditionally because of their diet. Methane is a greenhouse gas.

All large supermarkets now sell organic food. Many people will pay more for food which they think is safer, more nutritious and less damaging to the environment. Some people use farmers' markets or have boxes of local fruit and vegetables delivered to their homes.

Government policies

The Common Agricultural Policy (CAP)

Since the 1960s, the EU has given farmers subsidies through the Common Agricultural Policy, usually known as the CAP. In the past, the CAP encouraged farmers to produce as much food as possible. Now subsidies are linked to high-quality products, animal welfare and protecting the environment.

About €53 billion a year is distributed through the CAP – roughly 40% of the total EU budget. This percentage is falling and should be down to 33% by 2013.

There are about 12 million full-time farmers in the 27 EU countries eligible for support. Eighty per cent of the money goes to 20% of the farmers as they own most of the land.

Farmers can apply for a **Single Farm Payment** if they are producing high quality crops and have high standards of animal welfare. The amount received depends on the size of the farm; farmers with large farms receive larger grants.

Rural Development

Part of the CAP budget is reserved for Rural Development (see page 106) and protecting the environment.

The Environmental Stewardship Scheme (See page 106.)

The Energy Crops Scheme – This provides grants to encourage farmers to grow biofuel crops instead of food crops. The aim is to increase the amount of energy crops grown, and farmers can get up to half the start-up costs for producing fast-growing woody plants that provide fuel for power stations or biomass boilers. This renewable energy is used instead of burning fossil fuels, such as oil and gas, and is helping the UK to meet its target for renewable energy and to combat global warming.

Farming in tropical and sub-tropical areas

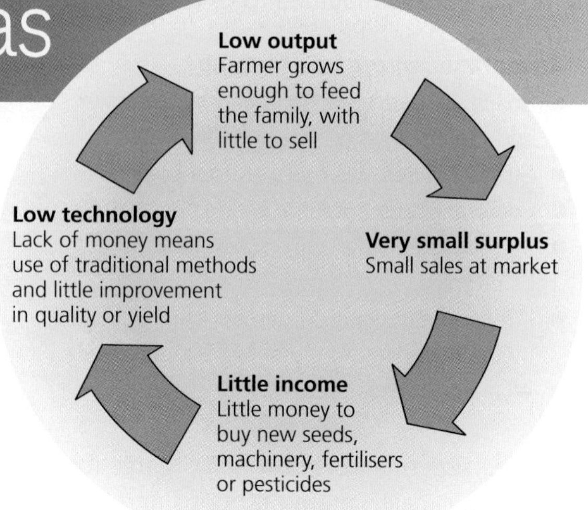

Subsistence farming

Most people in low-income countries are subsistence farmers. **Subsistence farming** means producing crops and rearing animals for the use of the farmer and his family. Only the surplus produce, if there is any, is sold.

Low output
Farmer grows enough to feed the family, with little to sell

Very small surplus
Small sales at market

Little income
Little money to buy new seeds, machinery, fertilisers or pesticides

Low technology
Lack of money means use of traditional methods and little improvement in quality or yield

What all types of subsistence farming have in common

Cash crops

In poorer countries subsistence farming is being replaced by **cash crops** such as bananas, sugar cane, cocoa, cotton, coffee, and cut flowers. These crops are not sold locally but are exported, often by transnational corporations (TNCs), to help the country earn money and pay off debts.

- Usually grown on a large scale as this is more efficient.
- Chemical fertilisers and pesticides are used to increase yields.
- Machinery may used, e.g. for spraying chemicals or harvesting, but this may be done by local people who are employed on a seasonal basis.
- Modern irrigation techniques may be introduced.
- Products are sold, often to a TNC, and exported.

Impact on subsistence economy

- Many subsistence farmers rent land from a large landowner; they may lose their land if large-scale cash cropping is introduced.
- Jobs on the farms are seasonal and for casual workers so there is no job security and people may have no income at certain times of the year.
- Increasing mechanisation means fewer jobs are available for local people.
- Lack of jobs encourages people to move to the city to find work – rural–urban migration.
- Chemicals used on the farms can pollute the environment.
- Cash crops are exported so food may need to be brought into the area.

Rural–urban migration

Changes to farming methods in poorer countries are causing **rural–urban migration** (see p.90).

Women and children may remain in the village while men go to find work in the town. They send money home to help the family. The women have to look after the farm and children, who also need to help out, and may have to give up going to school. Eventually the 'pull' of the town may encourage the whole family to move.

Soil erosion

Fertile top soil may be blown or washed away. Worldwide, an area ten times the size of the UK has become so badly eroded it can no longer be cultivated. Soil erosion is most likely to occur on steep slopes especially where there is heavy rainfall, e.g. in tropical areas.

Actions which increase the risk of soil erosion	Actions to prevent soil erosion
• Deforestation increases surface runoff and water may remove soil especially on steep slopes. • Removing hedges exposes soils to wind erosion. • Ploughing straight up and down a hillside increases surface runoff and gully erosion. • Overgrazing animals exposes land to erosion as there is not enough vegetation to bind soil. • Growing just one crop (monoculture), which is all harvested at the same time, can expose the soil. • Over-cropping removes organic matter and weakens the soil so erosion is more likely.	• Plant trees on steep slopes to reduce surface runoff. • Plant trees or hedges to provide shelter. • Plough across the slopes (contour ploughing) to slow surface runoff. • Reduce number of grazing animals. • Introduce other types of crops where monoculture is practised. • Lay magic stones – lines of stones built across the hillside to reduce soil erosion.

Irrigation

- **Irrigation** is adding water to farmland by artificial means.
- It is used in areas where there is not enough rain or the rainfall is unreliable or seasonal.
- Irrigation water is often stored behind a dam, e.g. the Aswan Dam on the River Nile.
- Water is taken to the fields in channels, using pressure sprinklers, or using drip irrigation.

Impacts of irrigation

- Irrigation increases yields so farmers have more to sell
- It is expensive to set up especially if a dam needs to be built to store water, although this can be used to generate hydroelectric power
- It can encourage farmers to grow cash crops resulting in tenant farmers being forced off the land
- Water-borne diseases can increase if irrigation water is channelled in open ditches through fields

Irrigated land is more than twice as productive as rain-fed land. About 16% of the world's farmland is irrigated, but this produces 36% of the world's food. By 2025, 80% of food is expected to come from irrigated land.

Salinisation

The most serious problem caused by irrigation is **salinisation**. Once soils become saline (salty), plants cannot grow and in some cases farmland has to be abandoned. Salinisation is a problem in many countries including Australia, USA, Egypt and Pakistan.

Appropriate technology

High technology equipment may be unsuitable for people in poorer countries who do not have the money or skills needed. **Appropriate technology** is equipment which is at the right level for local people to use and which uses local or cheap materials and is easy to maintain.

In Kenya, a simple foot operated pump is used by farmers to irrigate their land. The super money-maker pump was developed by an organisation called 'Kickstart'. It can draw water from a well, river or pond and pump it through a hosepipe onto crops or into a storage tank. It is cheap to buy and simple to operate and can be used to irrigate up to one hectare of land.

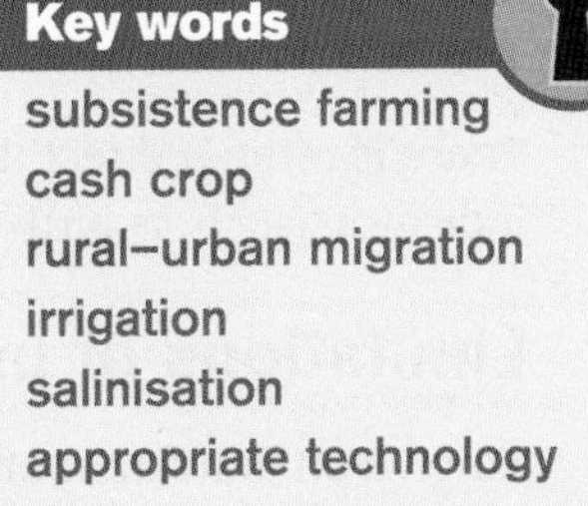

Key words
subsistence farming
cash crop
rural–urban migration
irrigation
salinisation
appropriate technology

Measuring development

Measuring development must take into account many factors apart from wealth. Development also depends on factors like education and health care. We would not say that a country was developed unless people living there were able to enjoy a reasonable quality of life.

The table shows seven **indicators of development** that can be used to try to decide how developed a country is. Other indicators, such as infant mortality rates or access to safe water, can also be used. The table shows big differences between the three countries.

Country	HDI rank	GDP (PPP US$)	Birth rate (per 1,000)	Life expectancy (years)	Urban population (%)	Adult literacy (%)	People per doctor
Nigeria	158	1,969	43	47	47	72	3,571
South Korea	26	24,801	10	79	82	97	1,076
UK	21	35,130	13	79	90	99	623

Measures of development

Gross Domestic Product (GDP) – is a measure of a country's wealth.

Advantage: a clear measure of economic development.

Disadvantages:

- Does not indicate social well-being or quality of life.
- Data easier to collect in rich countries than poorer ones where many people work for cash in the hidden economy and where methods of recording are unreliable.
- The figures give an average for the whole country and do not indicate the inequalities within the country.

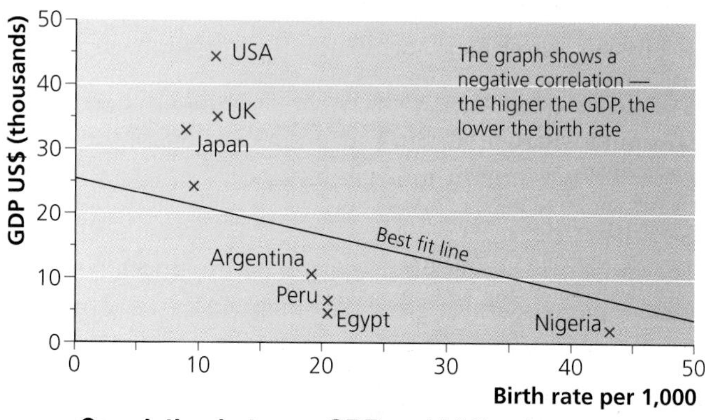

Correlation between GDP and birth rate

To overcome some of the disadvantages of GDP it is often better to use the **Gross National Income per capita (GNI)** as an indicator.

The **GNI** measures the average income of people in a country in US$.

There is often a direct link or **correlation** between wealth measured as GDP or GNI and other indicators such as birth rate and literacy.

Limitations of using one measure of development

To give an indication of the quality of life in a country, the United Nations developed the

Human Development Index (HDI) – to describe both economic and social well-being within countries. The HDI is based on three factors: life expectancy, literacy and GDP.

Advantage: the HDI is thought to give a better indication of the quality of life.

Disadvantage: it only includes three indicators, which may not necessarily be the best indicators. Should there be more indicators used to give a more accurate picture?

GDP and HDI can give very different results. For example in 2006, South Africa was ranked 70th in the world in GDP but 125th on the HDI table.

There is a clear difference between quality of life and **standard of living**.

Standard of living – refers to the money a person has for daily living and whether it is enough.

Quality of life – refers to what is important to a person's happiness such as having a job, education, health, friends, freedom, security, a good diet, water supply other considerations as well as income.

Different ways of classifying different parts of the world

In 1971, the influential Brandt Report divided the countries of the world into two categories: the rich industrialised countries mainly in the northern hemisphere, and the poorer, mainly agricultural countries of the southern hemisphere. The line dividing them became known as the 'north–south line'.

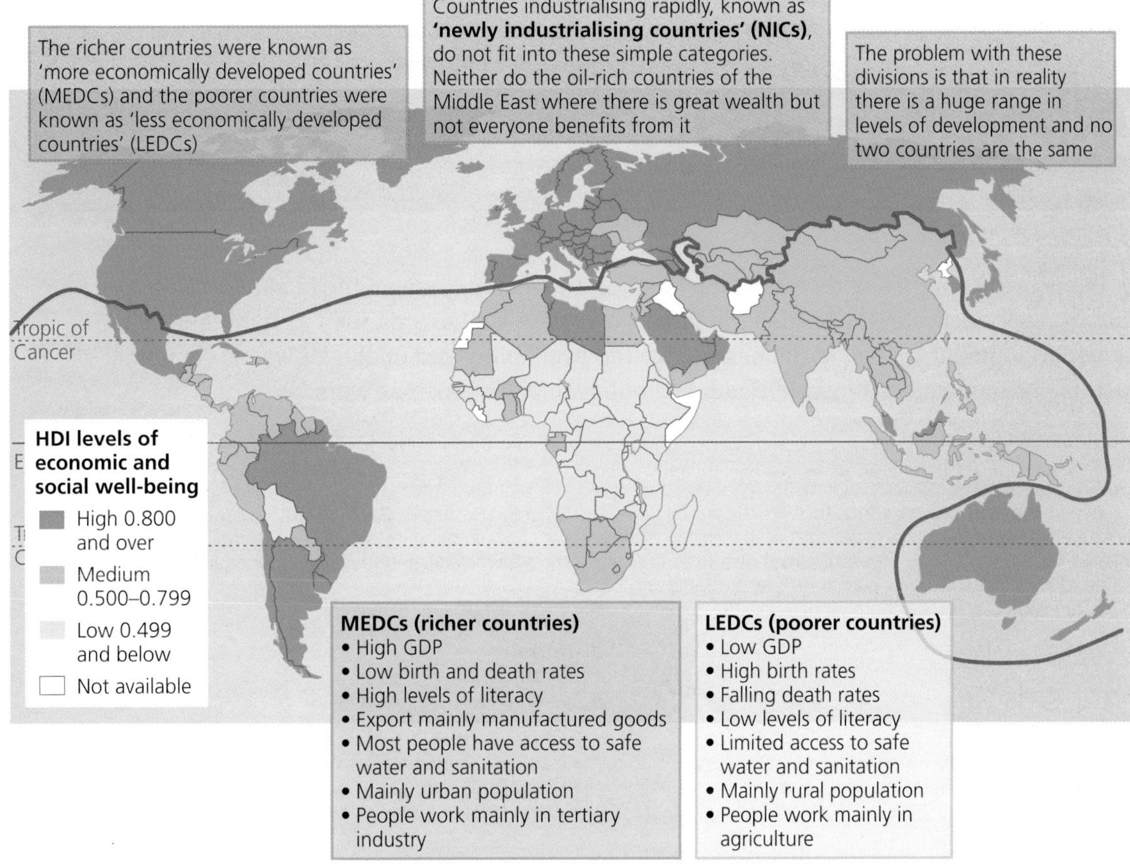

The richer countries were known as 'more economically developed countries' (MEDCs) and the poorer countries were known as 'less economically developed countries' (LEDCs)
Countries industrialising rapidly, known as **'newly industrialising countries' (NICs)**, do not fit into these simple categories. Neither do the oil-rich countries of the Middle East where there is great wealth but not everyone benefits from it
The problem with these divisions is that in reality there is a huge range in levels of development and no two countries are the same
Tropic of Cancer
HDI levels of economic and social well-being
High 0.800 and over
Medium 0.500–0.799
Low 0.499 and below
Not available
MEDCs (richer countries)
• High GDP
• Low birth and death rates
• High levels of literacy
• Export mainly manufactured goods
• Most people have access to safe water and sanitation
• Mainly urban population
• People work mainly in tertiary industry
LEDCs (poorer countries)
• Low GDP
• High birth rates
• Falling death rates
• Low levels of literacy
• Limited access to safe water and sanitation
• Mainly rural population
• People work mainly in agriculture

Case study: *people in poorer places*

Attempts to improve quality of life

Turkana, northern Kenya

The problem: The people are poor. They are pastoralists, herding livestock they move from place to place and the children help herd the animals. Children's attendance at school is poor so they receive little education. Teachers lack essential training.

Aim: To ensure that all children receive an appropriate education to help succeed and find a way out of poverty.

Solution: Aid through Oxfam provides and improves mobile schools to help all children attend school and give all children a decent education. Also to train teachers and offer support in order to improve the quality of education provided.

Rural Honduras

The problem: People lack the knowledge, skills or money to make the most out of their land. Help is needed to enable people to overcome poverty.

Aim: To help people grow enough food for themselves, improve their diet and to earn an income.

Solution: With the help of Oxfam, to teach sustainable farming methods and enhance their skills in order to reduce costs of purchasing seeds, fertiliser and pesticides. To help increase farmers' incomes by concentrating on crops which earn the highest prices. To establish schemes that provide loans for farmers to invest in their farm.

Global inequalities

In the poorer countries of the world the balance between improving the quality of life and failure is very delicate. Environmental factors and physical hazards such as hurricanes can slow development, as can social factors like the availability and quality of water, economic factors like inequalities in world trade, and political factors such as wars.

Factors that worsen global inequalities

Environmental factors
- Natural hazards — earthquakes, volcanoes, tsunami, and tropical revolving storms can all slow development. A poorer country is less able to cope with such hazards than a rich country
- Climate — many of the poorer countries lie in tropical regions. Diseases such as malaria and fevers weaken people so they are unable to work hard. Climatic events such as droughts and floods can wipe out years of hard work

Economic factors
- Poverty
- Unfair trade
- Processing — poorer countries export raw products which bring less profit
- Ownership — industry in poor countries by TNCs

Social factors
- Water supply — contaminated water supply causes diseases; time is spent fetching water and it is needed for irrigation

Political factors
- Corrupt governments — those in power take all the money
- Unstable government — industry won't invest
- Wars

Case study: *a climatic hazard event in a poor country*

Ethiopia, 2006

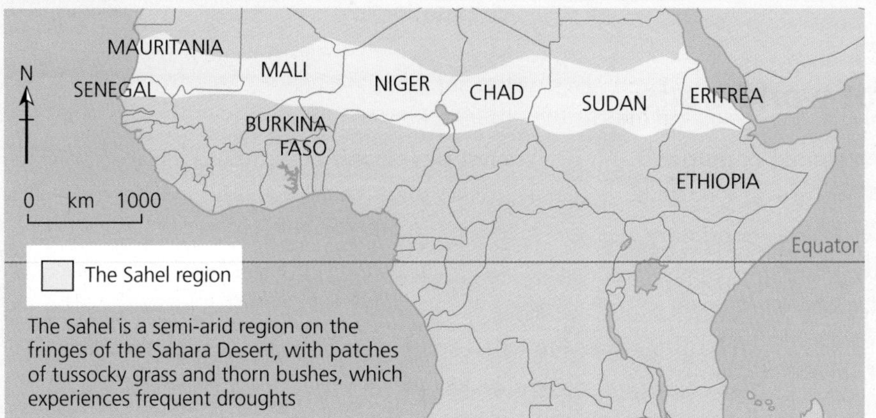

Ethiopia and the Sahel

Factfile

- **Population** – 77.7 million.
- **Area** – five times the size of the UK.
- **Wealth** – one of the poorest countries.
- **Climate** – seasonal rainfall.
- **Agriculture** – low technology.
- **Employment** – over 80% farming.
- **Civil war** – from 1998 to 2000.
- **Health** – only 20% of the population have access to safe drinking water.
- **Aid** – Ethiopia gets more relief aid than any other country.

Causes of drought

- Ethiopia does not have the infrastructure of dams and wells to cope with a drought.
- In the south and southeast of the country the recent rainy seasons were too little, too late and too erratic for water sources to replenish.

Impact of the drought

The short and long rains failed and caused a water shortage. The grazing for livestock dried up and much of the livestock died; water supply became critical. Crops failed, people became weak due to lack of food and water, and diseases like malaria and diarrhoea became common.

The response

The government turned to the world for aid.

- Food and drinking water provided by aid agencies were distributed at feeding shelters. Vaccines and emergency water tanks were sent to emergency stations.
- Aid also provided dry region plant seeds like maize and beans and tools such as hoes and sickles so that farmers could plant crops. New breeding stock for the livestock herders was provided so they could build their herds. Repair and maintenance of existing water points and advice on future water supply was also provided.

Long-term planning

Long-term planning is required to combat the effects of droughts and develop more sustainable support for the people enabling them to survive future droughts. It is important to develop low-cost systems to conserve water supplies, e.g. building low mud walls across slopes to slow water runoff; building micro-dams to store water; making sure existing systems are repaired; planting trees to conserve moisture; planting drought-resistant crops.

The imbalance in world trade

The pattern of world trade

The **poorer less developed countries** have only about a 20% share of world trade.

They tend to sell raw materials or **primary products** and to **import** manufactured goods. Primary products have less value than manufactured goods and the price is controlled by the richer countries' buyers.

The price of primary products often fluctuates on world markets. This can cause great difficulties for the producers. They cannot plan ahead as they don't know how much they will receive for their product. This is especially true if the country is dependent on one or two products only.

The **richer developed countries** produce about 70% of the world's **exports**, selling mainly to other richer countries, but also to poorer countries.

The newly industrialising countries (NICs) such as South Korea and Taiwan are also important trading nations.

Richer developed countries and NICs mostly export manufactured goods, which have a higher value than raw materials.

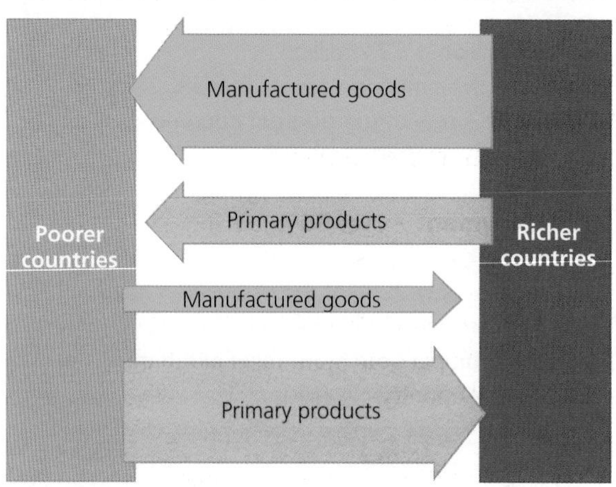

The trade balance between the richer and poorer countries

Unfair trade

Trading blocs – groups of countries often join together and make trade agreements to help each other (e.g. the EU).

Trade barriers – governments may use tariffs to stop too many imports into their country. Quotas may be used. These limit the amount of a product that may be imported. Governments use trade barriers to try to protect their own industries.

Debt

In the 1960s and 1970s, many poorer countries were persuaded to borrow money from richer countries. At the time interest rates were low.

During the past 30–40 years the repayment of debts has become increasingly difficult for many countries. They may owe much more money than they first borrowed because of the interest they have been charged.

Debts are stopping poor countries like Tanzania from developing. The government has to spend so much money repaying the country's debt that little is left for education and health services.

Reducing global inequality

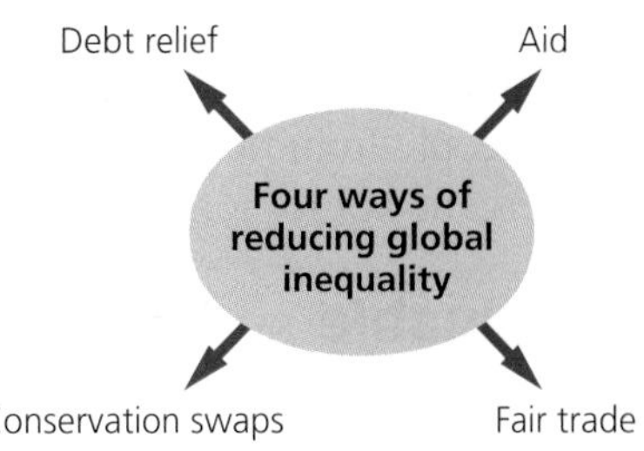

Debt relief

Debt relief is reducing the debt owed by poorer countries either by reducing the interest or the amount of the debt. Some of the richest nations, like those in the G7 economic group, of which the UK is a member, are promising to write off the debts of the poorest countries.

Conservation swaps

Swaps are agreed so that a richer country will write off part of a poorer country's debt if that poorer country will protect a special environmentally sensitive area. For instance, setting aside a part of the rainforest as a nature reserve and helping local communities with ecotourism. Over 50 countries now participate.

Case study: *fair trade*

Santa Anita la Union, Guatemala

Fair trade is an international movement that provides a guaranteed price for agricultural products over several years. This allows the producer to plan ahead.

The cooperative

- The Santa Anita cooperative produces organic coffee.
- Thirty-two families work as a cooperative.
- During the year they clear the land and weed round the bushes. It is four years before new bushes produce coffee beans.
- During the harvest the coffee beans are weighed to see how much each family has harvested, washed, processed and sold.
- In 2004, Santa Anita sold its coffee to Cooperative Coffees in the USA.

The future

The Fair Trade scheme has enabled this cooperative to develop their business, improve their standard of living and quality of life. Their children have been to good schools and are going on to college, using a bus paid for by the Fair Trade scheme.

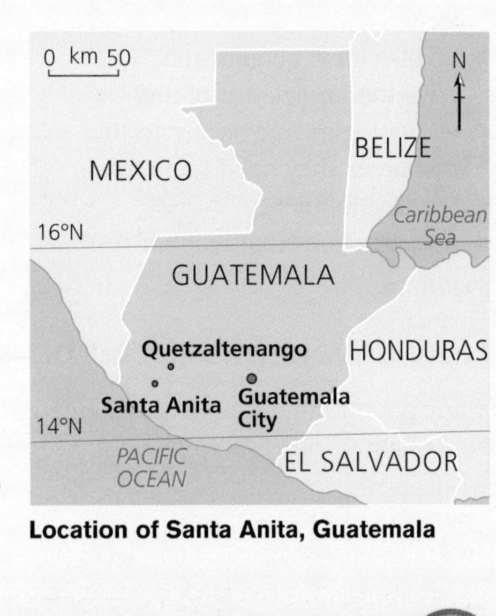

Location of Santa Anita, Guatemala

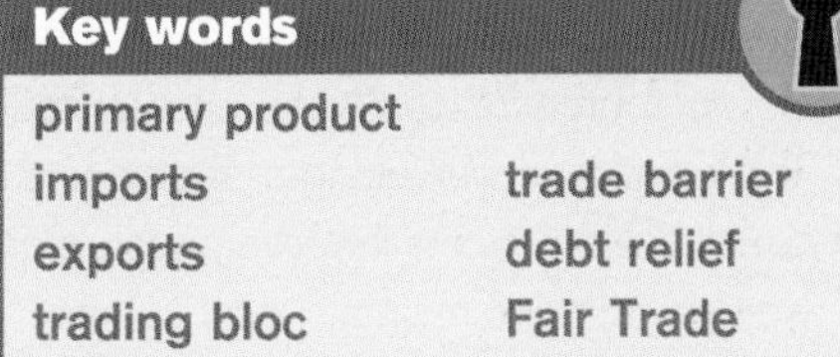

Key words

primary product
imports
exports
trading bloc
trade barrier
debt relief
Fair Trade

Aid

International aid is the transfer of resources such as money, equipment, food, training or skilled people from richer to poorer countries. Aid is given to help poorer countries either in an emergency or for long-term economic development.

Sustainable development

Sustainable development considers current needs as well as the needs of future generations. The earth and its resources should be handed over to the next generation in the same condition as they are in today. Sustainable development should:

- be able to continue in the long term
- involve local people who have the knowledge of the technologies and access to the resources they need to maintain the development.
- use resources that will not run out

Different types of aid are given in different ways

Bilateral aid
Aid given directly from one government to another in the form of money, training, technology, food or other supplies. In some cases this aid is tied aid, which means it has conditions attached that will usually benefit the donor country

Multilateral aid
Aid that comes from a number of different governments or organisations. It is usually arranged by an international organisation such as the World Bank or the United Nations (UN). These organisations usually give to large-scale projects

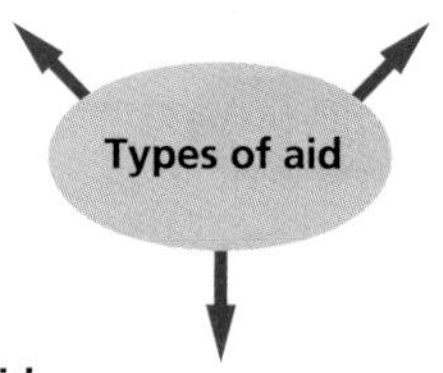

Non-governmental aid
Organisations such as Oxfam and Save the Children run projects all over the world, many of which are small-scale. They also help to organise **emergency aid** after disasters. These **non-governmental organisations (NGOs)** raise their money through donations and from government grants.

How development can be affected by aid

Many poorer countries have benefited from international aid, but their development may also be a result of other factors.

Benefits of aid

- Can be used to provide new technology and machinery that can produce more jobs, and enable farmers to grow more crops and increase exports.
- New infrastructure projects provide new roads and power sources. Clean water and sanitation improve the health and well-being of the people.
- Small-scale projects improve people's quality of life, self esteem and maintain local culture.

Shortcomings of aid

- Large capital-intensive projects may have unforeseen environmental and social consequences.
- The aid, particularly in the case of large projects, may not benefit the poorest people.
- Much foreign aid is bilateral aid, which is tied to joint projects and trade agreements.

Case study: *a small-scale development project*

Basket weaving: sustainable development in Ghana

Many aid organisations believe that small-scale, locally based projects are more sustainable than the large-scale projects.

1 The north of Ghana is mainly dry savannah land. The people are mainly subsistence farmers. The average income is below internationally recognised poverty levels with earnings of less than US$1 per day

2 The people had a long tradition of basket making using the native savannah grasses

3 With an aid grant from Oxfam villagers were offered training in business management and encouraged to develop their basket-making skills and set up small businesses

4 Within 5 years the basket weavers' network had 408 members — 330 women and 78 men

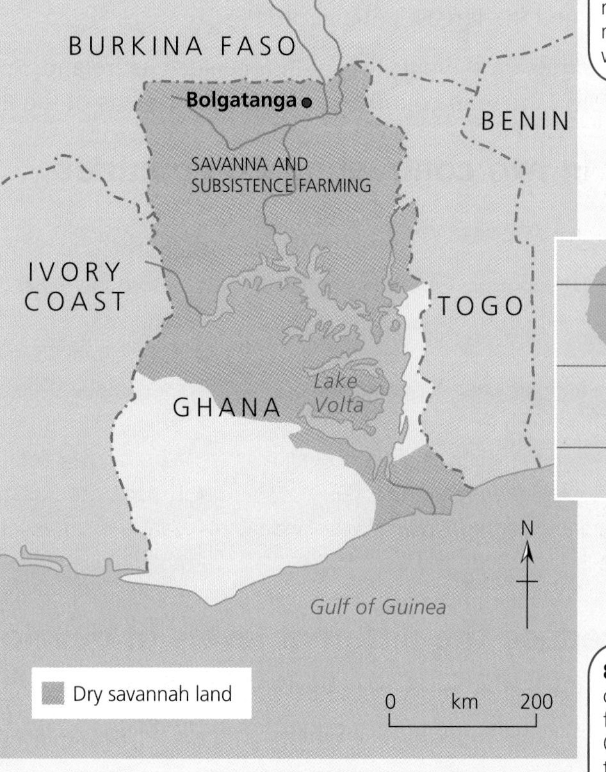

5 Changes in the environment meant that the right grass was becoming more difficult to find locally. It is brought from the south of the country to a central straw bank in Bolgatanga. Weavers collect straw on credit and pay back the loan when the baskets have been sold

6 The possibility of growing more of the right grass locally is being investigated

7 The baskets were sold in the local markets. However, many local people began to use plastic containers, and traders bought baskets cheaply and sold them in the markets elsewhere in Ghana. Income for the basket makers began to decline

8 The basket makers' organisation looked for other markets. Oxfam bought designs to sell in its shops in the UK. Other outlets have been found in the USA and Denmark

Key words

international aid
sustainable development
bilateral aid
multilateral aid
emergency aid
non-governmental organisation (NGO)

Development in the EU

Core and periphery in the European Union (EU)

Geographers have always recognised that there are great variations in the wealth and quality of life in member states.

The core – the heart of the EU is its wealthiest region with the most advanced industry, the best communications and the largest population.

The periphery – countries on the outskirts of Europe such as Ireland, Spain, Portugal and Greece and the eastern European countries are the poorer areas of the EU.

Development in two contrasting EU countries

Netherlands

- One of the original members of the EEC, it lies at the heart of the core region with the 16th largest economy in the world.
- A great trading nation and has been for centuries.
- At the mouth of the Rhine with Rotterdam one of the greatest ports in the world trans-shipping petrochemicals and general cargoes to the core area of Europe.
- Strong agricultural sector exporting flowers, bulbs, and greenhouse products like tomatoes and cucumbers.

Bulgaria

- Joined the EU in 2007 and lies on the periphery.
- Few natural advantages: a mainly rural country for centuries.
- Arable farming: sunflower seed, raspberries, tobacco, flax and chilli peppers.
- Exports raw materials such as iron, copper, coal and construction materials, electronics and petroleum fuels.

Attempts to reduce the different levels of development

The wealthiest states contribute most to the EU budget. About a third of this budget pays towards regional aid to benefit the poorer countries in the EU.

Structural Funds

European Regional Development Fund – promotes the creation of new jobs, improvements to **infrastructure** (including roads, railways, energy supply, water and sewerage), the activities of small- and medium-sized businesses and supports the most deprived areas of cities.

European Social Fund – aims to prevent unemployment and pays for education and training.

Cohesion Funds

Improve the environment and infrastructure in the less wealthy member states. Spain, Greece, Portugal and Ireland are the main countries that have benefited.

Key words

European Union	European Regional Development Fund
core	infrastructure
periphery	European Social Fund
Structural Fund	
Cohesion Fund	

Test yourself

1 **What is the main advantage of measuring development using HDI rather than GDP?**

2 **Name two measures of development used to determine HDI.**

3 **What is the difference between 'quality of life' and 'standard of living'?**

4 **List the factors that make global inequalities worse.**

5 **Match these terms to the correct definition:**

Bilateral aid	short-term immediate relief after a disaster
Tied aid	development projects run by charities such as Oxfam
NGO aid	aid given by one government to another
Emergency aid	aid arranged by international organisations, e.g. the World Bank
Multilateral aid	aid given with conditions, usually to benefit the donar

6 **List the ways in which the EU has attempted to reduce the differences in the level of development between countries within the EU.**

Examination question

Exam tip

To gain marks, always make sure that you write about a case study that you have studied if you are asked to in the later parts of the question.

Foundation tier:

(a) Describe how the development of one country that you have studied has been affected by a natural hazard.
- **(i) State the name of the country.**
- **(ii) Describe the type of natural hazard.**
- **(iii) Describe the ways in which the development of the country has been affected by the natural hazard.** *(4 marks)*

(b) For one sustainable development project in a poorer country that you have studied describe the benefits the project has brought to the people.
- **(i) State the name of project.**
- **(ii) Describe the benefits.** *(6 marks)*

Higher tier:

(c) For a country you have studied: describe how the development of the country has been affected by a natural hazard. *(4 marks)*

(d) For one sustainable development project in a poorer country that you have studied explain how the project has brought benefits to the people. *(6 marks)*

Globalisation

Globalisation is the production of goods and services on a worldwide scale to supply a global market leading to increasing interconnectedness and interdependence.

Reasons for globalisation

The components of goods will often be designed in one country and made in several different countries. This is the result of **globalisation**. The reasons for this are:

- **Lower wages** – wages in China and the Far East are lower than in countries such as the USA, Japan and the UK.
- **Improvements in communications** – the internet, telephone and fax enable the headquarters of a company, for example in the USA, to keep a close watch on production, design, and research and development in a factory it owns on the other side of the world.
- **Improvements in transport** – enormous container ships moving goods across the seas have reduced transport costs.
- **Free trade** – over the past 25 years trade barriers and tariffs between countries have fallen considerably.

Many large companies such as Hitachi, Toyota, Nike and Coca-Cola have moved their manufacturing operations to the Far East into countries like South Korea, Taiwan and China.

Case study: *globalisation*

The Apple iPod

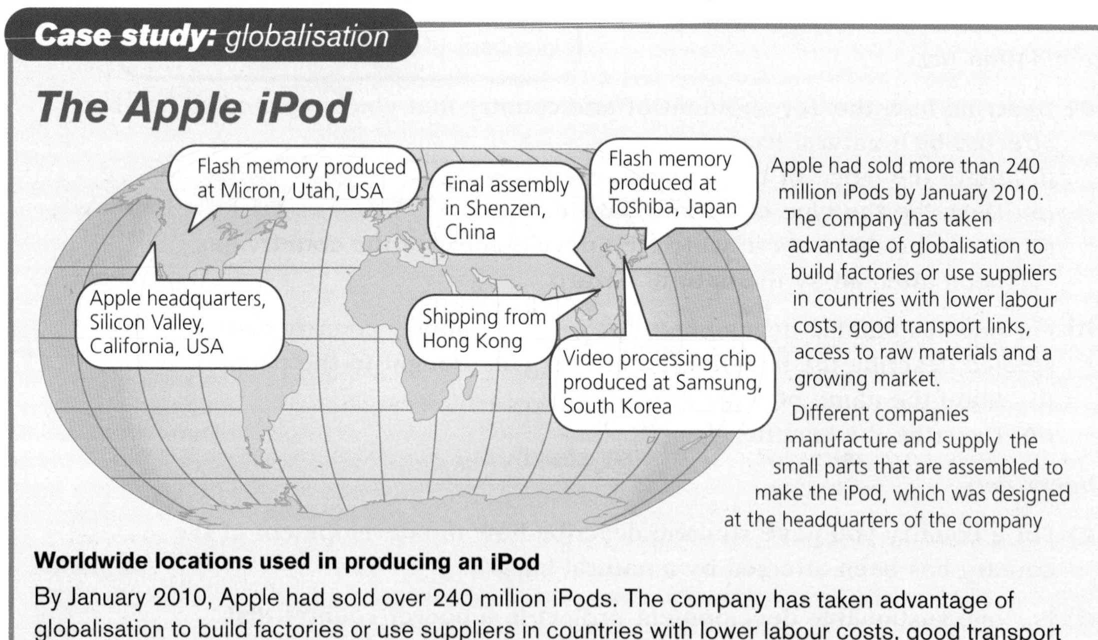
Flash memory produced at Micron, Utah, USA
Final assembly in Shenzen, China
Flash memory produced at Toshiba, Japan
Apple headquarters, Silicon Valley, California, USA
Shipping from Hong Kong
Video processing chip produced at Samsung, South Korea
Apple had sold more than 240 million iPods by January 2010. The company has taken advantage of globalisation to build factories or use suppliers in countries with lower labour costs, good transport links, access to raw materials and a growing market.
Different companies manufacture and supply the small parts that are assembled to make the iPod, which was designed at the headquarters of the company

Worldwide locations used in producing an iPod

By January 2010, Apple had sold over 240 million iPods. The company has taken advantage of globalisation to build factories or use suppliers in countries with lower labour costs, good transport links, access to raw materials and a growing market.

Different companies manufacture and supply the small parts that are assembled to make the iPod, which has been designed at the headquarters of the company.

Localised industrial regions with global connections

Silicon Valley, California, USA, is home to the headquarters of a host of ICT and other high-tech companies such as Apple Inc, eBay, Google, Hewlett-Packard, Yahoo, and Intel, because:

- similar companies are located nearby and knowledge can be shared
- education establishments are nearby to offer training to future employees
- there is a good climate, landscape and quality of life to attract the best workers
- these companies have global connections, e.g. Apple iPod

The development of call centres abroad

Companies locate call centres abroad in two ways:

1 **Outsourcing** – a larger company contracts some work to smaller specialist companies.
2 **Offshoring** – a large company develops part of its operation in another country.

Case study: *the development of call centres abroad*

Bangalore, India

The main reasons why Bangalore is a good location for large companies like BT to establish call centres are:

- Labour costs are much lower in India than in the UK or Europe.
- India has a large, skilled, English-speaking workforce, many of whom are university graduates.
- The Indian government supports the ICT industry and encourages investors to support new ICT industries.
- The rental cost of office and factory space is very low compared to that in richer countries.
- Bangalore stands 900 m above sea level, which gives it a pleasant climate for staff from Europe and America.
- As more ICT businesses grow in Bangalore, more workers are trained, knowledge can be exchanged, offshoot industries servicing the main companies can develop so the advantages of **agglomeration** can be utilised.

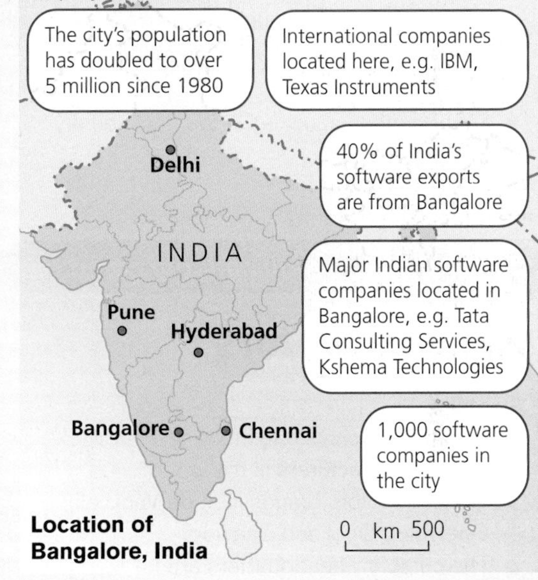

The city's population has doubled to over 5 million since 1980
International companies located here, e.g. IBM, Texas Instruments
40% of India's software exports are from Bangalore
Major Indian software companies located in Bangalore, e.g. Tata Consulting Services, Kshema Technologies
1,000 software companies in the city
Delhi
INDIA
Pune
Hyderabad
Bangalore
Chennai
0 km 500

Location of Bangalore, India

The benefits and issues of call-centre growth in Bangalore

Benefits	Issues
The creation of well-paid jobs in India Bangalore has developed as a city with new buildings The development of call centres has contributed to the Indian economy and some are helping the government to train people	People in Europe and America lose jobs Many people from the countryside have been attracted to Bangalore causing the city difficulties in servicing and housing due to the increase in population The benefits are largely restricted to the English-speaking middle classes in certain cities, mainly in the south

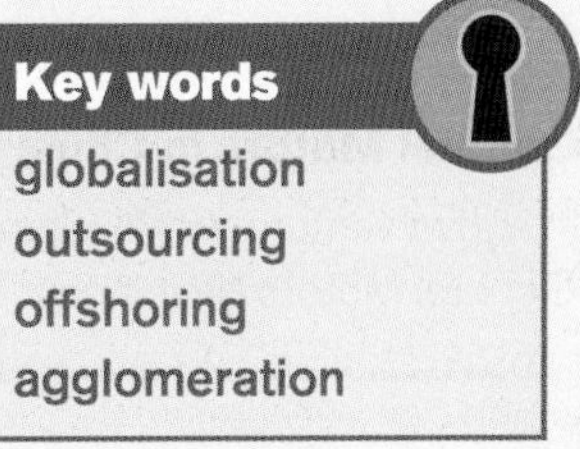
Key words
globalisation
outsourcing
offshoring
agglomeration

Transnational companies

Transnational companies (TNCs):

- are very large; some with an income higher than that of small countries
- trade across the world with their headquarters in one country (usually a richer country) and branch factories in many other, both richer and poorer, countries
- can benefit from owning factories in other countries – paying low wages, avoiding some taxes and tariffs, being near or inside a market

Advantages of TNCs

- Provide jobs and good wages.
- Provide training to improve skills.
- Develop new roads and services and bring investment into the country.
- Increase foreign trade and bring in foreign currency.
- Support other industries in the host country (the multiplier effect).

Disadvantages of TNCs

- Often locate headquarters, research and development in home country but pay low wages for manufacturing in other countries.
- Bring foreign nationals to fill higher-paid jobs.
- Make goods for export, not for the host country.
- Take profits out of the host country.
- In difficult times, close overseas factories first.

***Case study:** a multinational company*

General Motors

Factfile

- General Motors is one of the top ten motor manufacturers in the world.
- General Motors brands include Vauxhall, Chevrolet, Opel and Daewoo.
- The company headquarters are in Detroit, USA.
- General Motors operates in 32 countries worldwide.
- The company employs 266,000 people.
- GM operates factories in Ellesmere Port, Liverpool and Luton making Vauxhall cars.

General Motors in China

In a joint venture with a Chinese company, General Motors built a new car assembly plant to make 300,000 cars per year in Liuzhou in southeast China.

The advantages for General Motors are:

- The company can sell cars directly into the fast-growing Chinese market.
- China's rapid economic development means that there is an increasing demand for commercial vehicles and, as wages increase, for domestic cars.
- China has low wage costs so General Motors cars can be made more cheaply, increasing profits.
- Many electronic parts and steel are close by as they are already made in China.

The advantages for China are that training, skills and jobs are provided and factories supplying GM benefit.

Key words

transnational company (TNC)

Changes in manufacturing

In some countries industry is declining (de-industrialisation) while in others it is rising.

Reasons for the changes

Out-of-date methods Older, established heavy industries like iron and steel, shipbuilding and textile manufacturing in old factories in established areas like the UK could not compete with the modern methods being used elsewhere in the world.

Government legislation – regulating things such as the minimum wage to ensure that workers were paid a fair wage; health and safety making sure people were safe at work; limiting the number of hours worked, etc., all increased costs in some countries (usually the older, industrialised, richer countries). These regulations are not enforced in the newer industrialising countries.

Governments offer tax and trade incentives to attract investment to build new factories. Legislation to reduce the number of strikes makes the country more attractive to investment.

Globalisation The reduction of transport costs (using container shipping in particular) caused competition to become global. Stable countries with a good level of education and skills but low wages were able to attract industry and their economies grew.

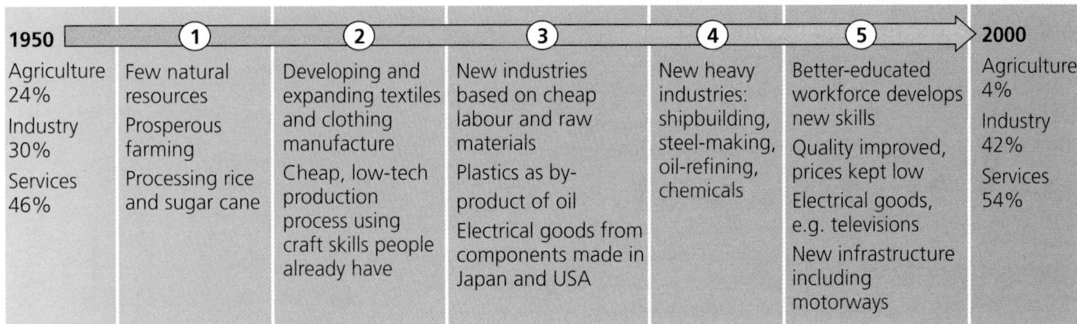

1950
Agriculture 24%
Industry 30%
Services 46%
1
Few natural resources
Prosperous farming
Processing rice and sugar cane
2
Developing and expanding textiles and clothing manufacture
Cheap, low-tech production process using craft skills people already have
3
New industries based on cheap labour and raw materials
Plastics as by-product of oil
Electrical goods from components made in Japan and USA
4
New heavy industries: shipbuilding, steel-making, oil-refining, chemicals
5
Better-educated workforce develops new skills
Quality improved, prices kept low
Electrical goods, e.g. televisions
New infrastructure including motorways
2000
Agriculture 4%
Industry 42%
Services 54%

The growth of industry in Taiwan

Examples of **NICs** are the southeast Asian countries of South Korea, Taiwan, Singapore and Hong Kong. They have broken out of the cycle of low profits and lack of investment by:

- investment from other countries, mainly Japan and USA
- investment in infrastructure (road, rail and ports) for communications
- early industries such as clothing, shoes, and paper products which relied on relatively cheap raw materials, abundance of cheap labour, low technology, and the use of craft skills already possessed by the workforce
- over time, the workforce developed new skills to make electrical goods, televisions, computers and cars

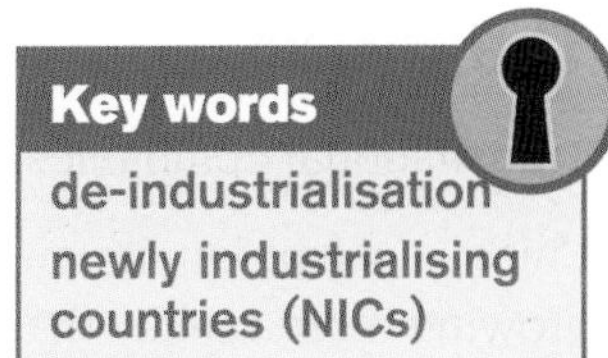

China – the new economic giant

Reasons for China's rapid growth

Government legislation

China realised it needed foreign investment and company expertise:

- foreign investment was encouraged
- in six areas, special economic zones (SEZs) were set up with tax incentives for foreign investors
- 'open' cities were declared, e.g. Shanghai
- the one-child policy checked population growth. People had more money to buy consumer goods like cars and televisions

Size of the home market

China potentially has a massive home market and the average income has more than tripled in the last 20 years.

Labour

China has an almost limitless source of labour, educational standards are good and wages are low.

Competitiveness

No other country can compete on cost with China, especially making clothes and toys.

Factfile

	1978	2008
Population	96.3 million	1,238 million
GDP per capita	US$ 56	US$ 3,332
Length of road	960,000 km	3,730,000 km
Value of international trade	US$ 1.13bn	US$ 2,561.6bn

China now has the third largest economy in the world.

Map of industry in China

Globalisation

The opening up of world trade, the lowering of trade barriers; the ICT revolution; and the reduction of transport costs all enabled China to trade with the world.

The result

Transnational companies were quick to establish factories in China.

China is expanding its own industries, selling to the world and its own people. The economy is growing more rapidly than any other country.

The growth of China's industries is taking work from traditional industrial areas in Europe that are unable to compete with the low production costs and wages.

Increasing demand for energy

Growing population and wealth

The reasons for the increasing demand for energy are interlinked. The world population is rising and as incomes increase demand for new technologies to improve quality of life also increases. New technologies need increasing amounts of energy. This is true in both the richer and poorer areas of the world.

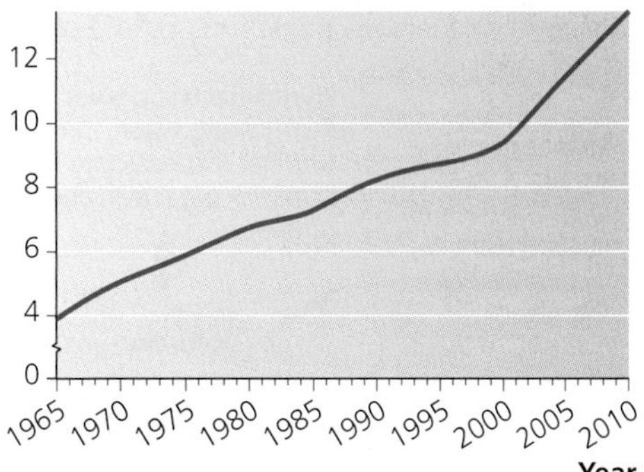

World energy consumption, 1965–2010 (million tonnes oil equivalent)

World population growth

The world's population is rising at a very rapid rate due to increasing food supplies and improved health care.

Increased wealth

People's incomes across the world are rising. In the richer countries this process has been going on for 250 years since the start of the Industrial Revolution. In the poorer countries in Asia, Africa and South America this process only started in the second half of the 20th century. With increasing wealth comes a greater use of technology to generate more wealth and provide a better standard of living. This leads to a greater use of energy.

Technological advances

As new technologies were developed during the last 100 years, the range of manufactured goods like washing machines, radio, television and cars has increased. As people buy more consumer goods the demand for energy grows.

Today people throughout the world work in factories and offices that require vast quantities of power to drive the machinery and for heat and light. They travel to work in cars or on trains or buses. All demand energy.

The impact of increased global energy use

Increased energy use

Social
Increasing use of energy has enabled people to have an improved quality of life.

- They have more leisure time as technology in the home reduces the workload.
- Their leisure activities change, travel for holidays is swift and easy, travel to newer types of entertainment is possible, e.g. cinema, shows and concerts.
- They are able to choose the environment in which they live and travel some distance to work each day.

Environmental
The greatest impact of increased energy use has been on the environment. Most of the energy we use comes from burning fossil fuels like oil, coal, gas and biomass. This damages the environment in the following ways:

- they release carbon dioxide which contributes to global warming
- mining causes environmental damage especially waste tips and scars left by open-cast mining
- pollution from oil spills from tankers and oil wells causes destruction of wildlife and landscapes, particularly coasts
- air pollution causes health problems

Sustainable development

Renewable energy – energy using a renewable resource such as water, wind or the heat of the sun, which unlike fossil fuels will never run out and doesn't pollute the environment.

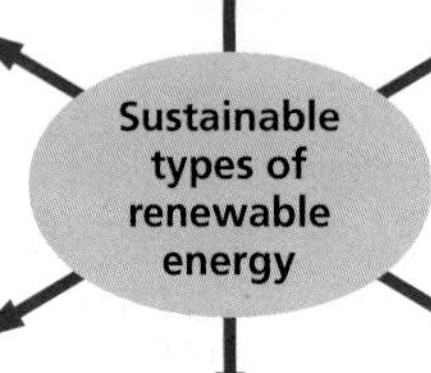

Hydroelectric power – fast-flowing water turns turbines. Expensive to build, cheap to run, clean cheap power

Nuclear power – uses uranium to heat water. Dangerous if radiation leaks so less popular than others, very expensive to build and to shut down

Wind power – large turbines with sails turned by the wind. Many needed to produce much power. Cheap to run, some people think are visually polluting

Geothermal power – (from water heated underground by volcanic action). Clean and efficient, only available in geothermally active areas

Solar power – at present mainly used on a small scale in individual homes. Cheap to run, costly to install

Tidal power – turbines in an estuary with a large tidal range. Few sites suitable, expensive to build. Cheap to run, produce large amounts of electricity, affect wildlife habitats

Renewable energy sources

- Only account for a small part of the electrical energy produced worldwide
- Are cleaner than non-renewable energy sources
- Are environmentally friendly, particularly in relation to air pollution
- Are expensive to develop but cheap to run
- Are sustainable – they will not run out
- Have potential for the future as technology develops

***Case study:** one type of renewable energy*

Wind farms in the UK

To meet its targets to reduce polluting carbon emissions, the UK government is focusing on nuclear power and wind power.

The 'go ahead' has been given to expand the number of wind farms on land but it is in the sea that giant wind farms are to be built with turbines 120 m high and as far as 18 km offshore.

Advantages of wind power
- No air pollution or carbon emissions.
- Cheap to run and quiet.
- No fuel to run out.

Disadvantages of wind power
- Visual impact on upland sites of high landscape quality.
- Wind is not constant and may stop blowing.
- Large numbers of turbines needed to produce a significant amount of power.

Advantages of offshore wind farms
- Not subject to planning laws.
- No cost of buying land.
- Turbines can be very large without being visually polluting.

Disadvantages of offshore windfarms
- Impact on birds migrating or moving to feeding grounds.
- Impact on radar systems.

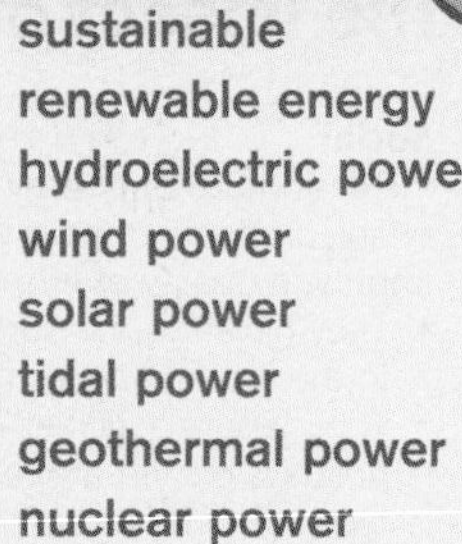

Key words
sustainable
renewable energy
hydroelectric power
wind power
solar power
tidal power
geothermal power
nuclear power

The global search for food

Globalisation has enabled supermarkets to source their food products from around the world. Wal-Mart (trades as Asda in the UK) and Tesco, the two largest multinational supermarket chains in the world, not only have stores in many different countries, but also source their food from across the world.

The benefits for the consumer are that they can buy their favourite vegetable or fruit all the year round even when it would be out of season in the UK.

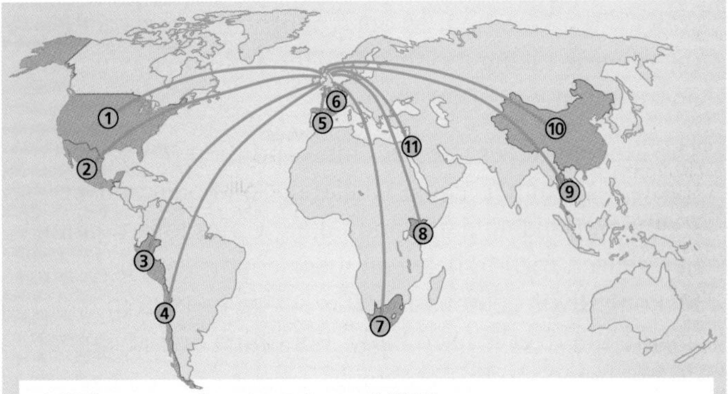

1 USA	Sweet potatoes – 5,900 km
2 Mexico	Red spring onions – 8,800 km
3 Peru	Asparagus – 10,000 km
4 Chile	Cherries, blueberries – 11,500 km
5 Spain	Raspberries, strawberries, fennel, lettuces, tomatoes, cucumber, aubergines, broccoli, marrows – 1,300 km
6 France	Mushrooms, butternut squash, potatoes – 300 km
7 South Africa	Grapes – 8,900 km
8 Kenya	Baby leeks, mange touts – 6,700 km
9 Thailand	Baby sweetcorn – 9,400 km
10 China	Ginger – 8,200 km
11 Israel	Radishes – 3,500 km

Distance travelled by some produce to one UK supermarket

Environmental issues

Carbon footprint – this is the amount of carbon created by the activities of people. For example, transporting food over large distances increases the '**food miles**' (the miles travelled by the food we buy) it travels. This will have a polluting effect on the environment from aircraft, ships, trains or lorries and increases the carbon footprint.

However:

- Runner beans in Kenya are grown without using the machinery, fertilisers and pesticides that would be used in the UK, so their carbon footprint is lower despite the need to fly and transport the beans overland to the supermarket.
- Cucumbers, tomatoes and other salad crops brought from Spain during the colder months in the UK may well have a lower carbon footprint than those grown in greenhouses in the UK at that time of year because they use large amounts of energy (heating and lighting) to produce. In the summer, when salad crops can be grown outside in the UK, the position is reversed and UK crops would have a lower carbon footprint.

Environmental degradation

As the population in some parts of the world increases, it creates extra pressures on the land to grow enough food to feed the additional people. This leads to increasing use of marginal land, which in the past would have been considered too poor to farm.

Clearing such land may lead to soil erosion and **degradation** (the decline in fertility) will result. The soil may be blown or washed away, or it may be overgrazed but the result will be the same – it will be useless for producing food.

Political issues

Often an increase in food production depends on a ready supply of water for irrigation. When a river flows through more than one country, disputes about water often break out as one country may feel that the other is taking too much water and not leaving enough in the river for them.

The Mekong River rises in China and then flows through Myanmar, Laos, Thailand, Cambodia and Vietnam. Building dams and taking water for irrigation has caused friction between these countries. The Mekong River Commission was set up in 1995 to ensure the river is used fairly. But China and Myanmar do not belong to this grouping.

China has started a programme of dam building on the river. The Mekong River Commission is concerned about the impact this will have on Cambodia where the river provides food and a livelihood for many.

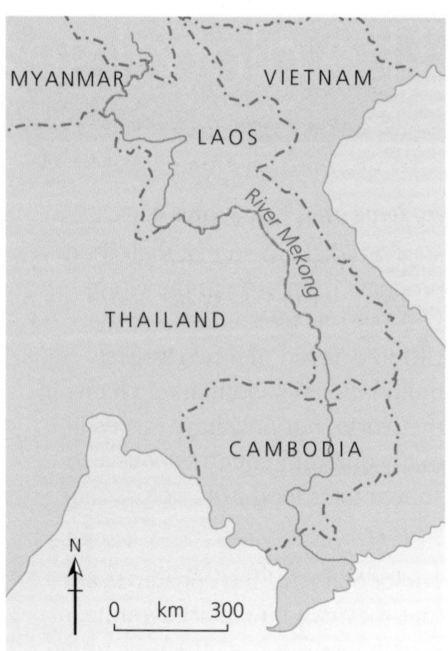

The Mekong River

Economic issues

Commercial farming to satisfy the demands of large supermarkets in richer parts of the world does not benefit local farmers except in providing some low-grade jobs. **Subsistence farmers** (producing just enough for their own needs and a little to sell) cannot increase their yield without buying higher yielding seeds, fertilisers and chemicals, so they must borrow money and go into debt. They are trapped in a cycle of poverty.

Social issues

Commercial farming (growing crops for cash) causes the breakdown of traditional ways of life. People who were farmers and grew their own food now work for someone else. If local farmers get into debt and become trapped in a cycle of poverty their only option may be to sell their land and move to the city.

Locally produced food

Increasingly people in the UK like to know where their food comes from and how it is grown. They are also concerned about food miles. There has been a rise in the number of people wishing to eat locally produced food.

Two methods of buying local produce have grown in popularity:

1 **Farmers' markets** – farmers and growers from an area sell their produce to the consumers at a market.

2 **Direct marketing** – local farmers and growers deliver produce to the customer's door often as a fruit or vegetable box.

Key words

carbon footprint
food miles
degradation
subsistence farmers
commercial farming
farmers' markets
direct marketing

Test yourself

1 (a) List the reasons for the growth of call centres in countries like India.
(b) Give *three* benefits of call centre growth for the host country such as India.

2 Give *three* reasons for the success of newly industrialising countries (NICs).

3 (a) What is meant by 'renewable energy'?
(b) Name *five* types of renewable energy.

4 Give *one* example of the international response to pollution control and carbon reduction.

5 What is meant by
- 'food miles'
- 'carbon footprint'
- 'degradation'
- 'subsistence farming'?

Exam tip

Make sure you learn the case studies required thoroughly so that you can give detailed answers with examples to gain full marks.

Examination question

Foundation tier:

(a) Give *two* reasons for the increasing global demand for energy. *(2 marks)*

(b) What are the environmental impacts of the increasing global demand for energy? *(3 marks)*

(c) For one type of renewable energy you have studied describe its advantages and disadvantages as an energy source. *(5 marks)*

Higher tier:

(d) Describe the major environmental impacts of the increasing global demand for energy. *(4 marks)*

(e) For one type of renewable energy you have studied describe its advantages and disadvantages as a sustainable energy source. *(6 marks)*

The growth of tourism

About 10% of the world's workforce is employed in the tourist industry, which generates a vast amount of money.

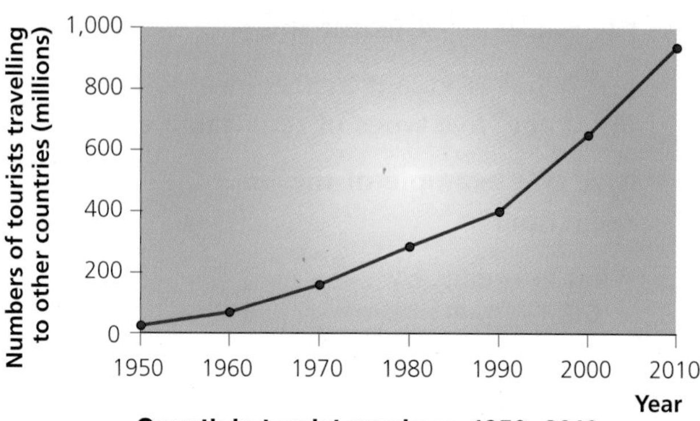

Growth in tourist numbers, 1950–2010

The graph shows how the numbers of international tourists have increased during the last 50 years and these numbers are expected to go on rising. There are several reasons for this.

- Transport – quicker and easier road and air travel developed in the 20th century.
- People in the richer countries are earning more, so they take more holidays.
- Many people, especially in richer countries have more days of paid leave.
- Travel companies provide a huge range of **package holidays** and other organised trips making travel much easier.
- People know much more about different places in the world through education, television programmes and the internet, and this encourages them to travel.

The potential of different environments for tourism

People like to go to somewhere different for holidays. This may be to a country with a different culture to their own or perhaps to experience a new environment.

Traditionally three types of environment can be identified:

Cities

- Historic attractions – palaces, cathedrals, castles and monuments.
- Dramatic structures such as the Sydney Opera House and the Eiffel Tower.
- General attractions, for example shopping, theatres and nightlife.
- Attractions unique to the city such as the canals in Venice and the skyscrapers in New York.

Mountains

- Snow sports such as skiing, snowboarding and tobogganing.
- Spectacular scenery with views of mountains, glaciers and lakes.
- Adventurous sports such as climbing, hiking and paragliding.

Coasts

- Most people travel to coastal resorts to enjoy the sand and sea.
- Cool weather and rain in the UK, even in summer, has encouraged people to look to the Mediterranean coasts as the nearest area with almost guaranteed sunshine, sand and sea.
- In winter, beaches further afield are either in the southern hemisphere, which are enjoying their summer, or the Caribbean, and Red Sea resorts nearer the equator are also popular.

The economic importance of tourism to different countries

Tourism in the UK

Thirty to forty million UK residents holiday in the UK each year. Most visit the coast, particularly the southwest coastal resorts with their warmer weather and higher number of hours of sunshine.

The numbers taking their holidays inland in areas of scenic beauty has increased significantly during the last 50 years in areas such as the National Parks and Scotland.

The importance of tourism

- Creates jobs and helps a country to develop industries such as construction and power
- Makes a huge contribution to government income and foreign exchange currency
- Creates jobs indirectly in civil service, e.g. in administration, police and health care
- Attracts investment to improve the infrastructure — roads, airports, water and power
- Provides opportunities for local industries such as farming and fishing to supply produce to tourists and shops, and handicraft industries to provide goods
- Is labour intensive, so creates a large number of jobs in the service sector directly serving needs, e.g. accommodation, food, entertainment and facilities for tourists

Visitors to the UK

Thirty million overseas tourists visit the UK each year. They contribute £15–20 billion to the UK economy.

Why people visit the UK	Factors influencing the number of people visiting the UK
• Countryside scenery • Sights of London • History and historic houses • Many visitors have ancestors from the UK	• The amount of publicity in their country about the attractions on offer in the UK • The strength of the pound against their currency – if the pound is strong then their holiday will be more costly • Terrorist threat – fewer tourists will visit if there is a threat of terrorism • The world economy affects the amount of money people have and if it is strong then they will be more optimistic and willing to come to the UK

The tourist area/resort life-cycle model

Tourist areas and resorts tend to have a pattern of popularity. From small beginnings they increase in popularity and then, after reaching a peak, numbers begin to decline and the resort or area has to work hard and make changes to maintain the number of visitors.

Tourist area/resort life-cycle model

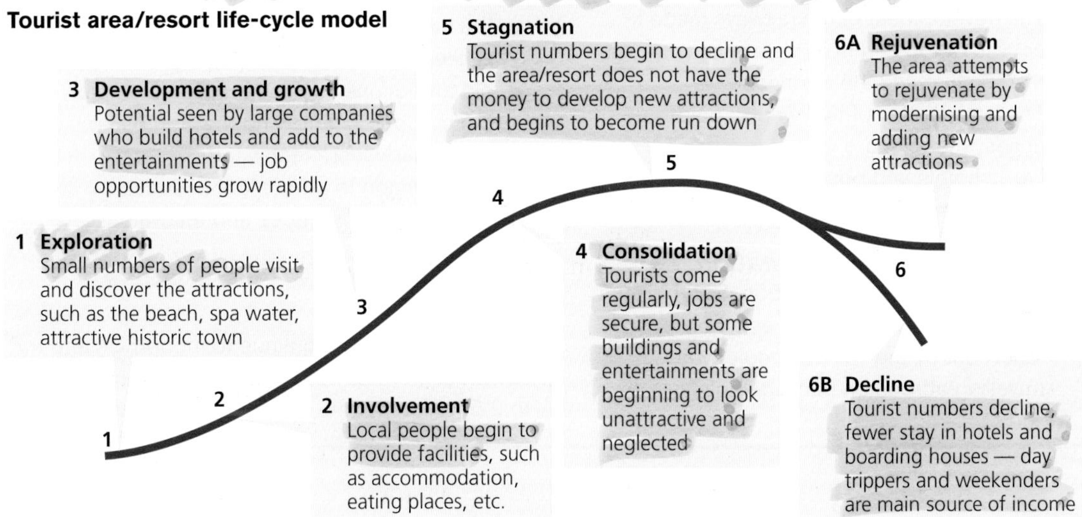

5 Stagnation
Tourist numbers begin to decline and the area/resort does not have the money to develop new attractions, and begins to become run down
6A Rejuvenation
The area attempts to rejuvenate by modernising and adding new attractions
3 Development and growth
Potential seen by large companies who build hotels and add to the entertainments — job opportunities grow rapidly
1 Exploration
Small numbers of people visit and discover the attractions, such as the beach, spa water, attractive historic town
4 Consolidation
Tourists come regularly, jobs are secure, but some buildings and entertainments are beginning to look unattractive and neglected
2 Involvement
Local people begin to provide facilities, such as accommodation, eating places, etc.
6B Decline
Tourist numbers decline, fewer stay in hotels and boarding houses — day trippers and weekenders are main source of income
1
2
3
4
5
6

Case study: a UK coastal resort

Scarborough

Factfile

Britain's first holiday resort with the advantages of:

- two long sandy beaches, the castle, spectacular cliff views and spa waters
- its position close to the large industrial towns of Yorkshire
- being very close to the North York Moors National Park

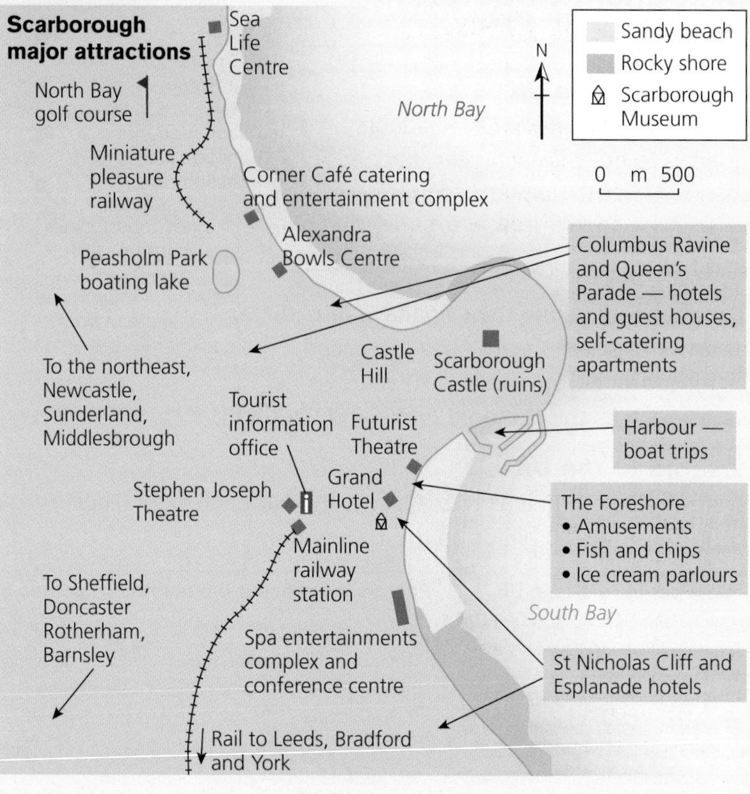

Beginnings

- The discovery in 1626 of mineral springs attracted visitors seeking the benefits of 'the waters'.
- By 1700 'people of good fashion' were also bathing in the sea.
- Local people were providing places to stay and restaurants.

Development and growth

- Major attractions were added such as the Grand Hotel and the Spa entertainment complex.
- The development of the railways made it easy to get to and from the industrial towns of Yorkshire and the northeast.
- The introduction of two weeks' paid holiday meant more visitors.
- Large hotels and boarding houses built.
- Entertainments opened along the 'front'.

Consolidation and stagnation

- From the 1930s to the 1950s, Scarborough maintained its importance as a holiday resort but by the 1970s people were travelling abroad on package holidays.
- As fewer people came to Scarborough, hotels became run down and attractions closed.

Rejuvenation

To halt the decline:

- the beach and water quality was improved. A new sea wall was built along North Bay
- old attractions such as the Spa Complex, the outdoor theatre, and the South Bay promenade were renovated
- efforts made to attract 'out of season' visitors to conferences and festivals

The future – were the strategies successful?

Scarborough must continue to modernise and try to attract more day trippers and weekenders. Up to 2.5 million people visit the resort each year.

Case study: *a UK National Park*

The Peak District

Factfile

- **National Parks** are protected areas of countryside protected so they can be enjoyed now and in the future.
- The Peak District was the first National Park in 1951.
- About 30 million visits are made to the park each year.

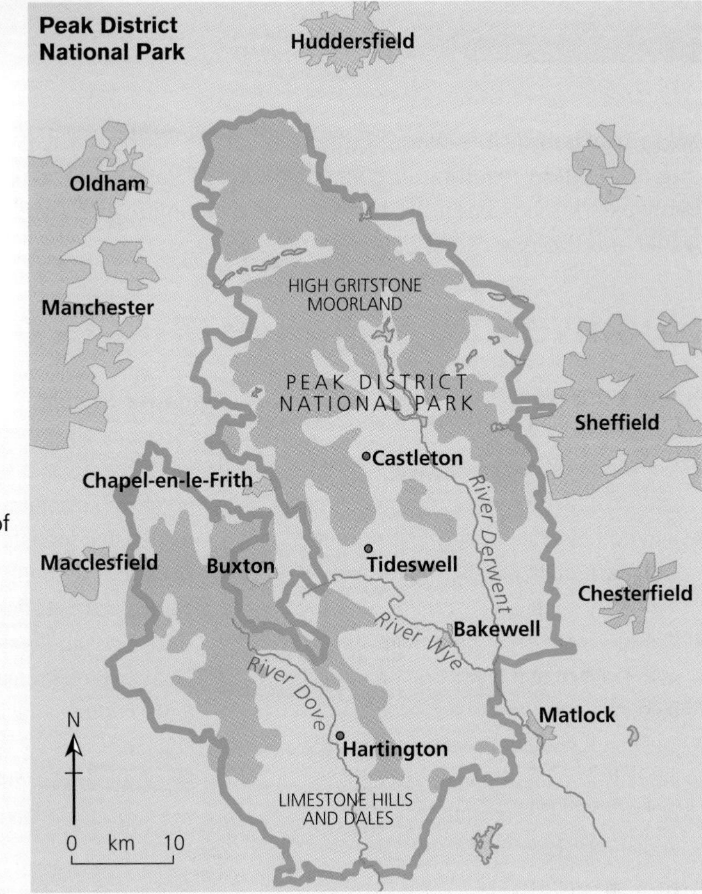

Why so many people visit

- Stunning scenery: high bleak moorland of gritstone and hills of limestone with grassland farming.
- Surrounded by motorways, the park lies within an easy drive of major cities.
- People come to walk, visit the villages and stately homes, and do climbing, caving, mountain biking and fishing.

Management issues and solutions

Issue	Problem	Solution
Honeypot sites	Many people visit the same attraction such as the pretty village of Castleton with its show caves, castle, shops and cafés. This causes congestion	The Peak Park Authority has built a large car park and visitor centre. Clear access footpaths separate people and cars
Traffic congestion	Almost 90% of the visits to the park are made by car. Congestion at peak times and parking are major problems with people parking in village streets and on verges causing more congestion and erosion	Designated touring routes are advertised and car parks are provided. Some roads have been closed and park-and-ride schemes have been developed
Litter	Inevitably with so many visitors litter will be dropped	Litter bins at all honeypot sites
Footpath erosion	So many walkers visit the park that the footpaths can quickly become bare and rainfall can erode deep gullies. Visitors may disturb sheep and cattle in fields	Many paths in the Peak District have slabs placed across them to stop further erosion, footpath signs mark routes clearly
Property prices	Property prices are high in such desirable areas and many local people are unable to afford their own home. In some villages, 60% of houses are second homes	Authorities build homes that are only sold to local people who live and work in the park

Key words

National Park
honeypot sites

Mass tourism

Mass tourism occurs where large numbers of tourists travel to one destination resulting in concentrations of hotels and tourist facilities. This has been the result of the availability and ease of air travel and the rise of the package tour company.

Key words

Mass tourism

Positive and negative effects of mass tourism

Positive effects

Economic

- Earns a great deal of money for the economy and foreign exchange currency.
- Improvements in the local infrastructure with new roads, airports, water and electricity supplies.
- Provides new job opportunities in hotels and facilities and a wide range of better paid jobs.
- New opportunities for the construction industry and local industries supplying the hotels and facilities with local produce and services.

Environmental

- Mass tourism may bring an awareness of the need to conserve the environment, not least for the tourists to enjoy.
- Tourism brings the money to develop schemes to conserve the environment.

Negative effects

Economic

- Holidays are sold by package holiday companies and hotels are owned by foreign companies so much of the money that tourists spend may end up abroad.
- Many jobs in tourism are poorly paid and low status, e.g. cleaners and waiters. Higher status jobs, such as guides requiring language skills, may go to outsiders.
- Tourist numbers may fluctuate and the area's popularity may change as fashion and economic circumstances change.
- Numbers of tourists and therefore jobs may be seasonal, which gives no job security.

Environmental

- Building more roads, airports, hotels and facilities may destroy the environmental quality of the area.
- Pollution may reduce water quality of the sea, and the area may have difficulty disposing of litter and waste.
- Local difficulties such as the reduction in fish catch, destruction of coral reefs, and landscape degradation.

***Case study:** an established tropical tourist area*

Phuket, Thailand

Factfile

- 11–12 million tourists visit the country each year – 55% from Asia (Malaysia, Japan and China) and 25–30% from Europe.
- Tourist numbers have fluctuated following the scares of the tsunami of 2004, bird flu, international terrorism and unrest in the capital Bangkok.

Why do tourists travel to Thailand?

- Rising costs of European destinations makes Thailand more affordable.
- Increasingly tourists, particularly from Europe, are looking for new exotic destinations and new cultures to enjoy.
- Thailand has Western-style resorts, sandy beaches, tropical warm weather and historic buildings.

The positive and negative effects of tourism

Positive effects

- Tourism has played a significant part in the economic development of Thailand.
- Tourism is Thailand's top earner of foreign currency.
- Tourism is a major employer because it is labour intensive

Negative effects

- Phuket favours large hotel chains owned by foreign companies who take profits out of the country.
- The quality of sea water is threatened due to dumping of waste, and coral reefs have been destroyed.
- The physical attractiveness of areas along the sea front has been destroyed by building modern hotels.

Strategies to reduce the negative effects of tourism

- Thailand advertises widely; one campaign was 'The land of smiles' based on its naturally welcoming people and their ready smile.
- Thailand has improved its infrastructure – its roads, water and sanitation – in resort areas to attract more visitors.
- Protecting areas of environmental value such as rainforest in the north of the island and Nai Yang Beach in the northwest where sea turtles lay their eggs.

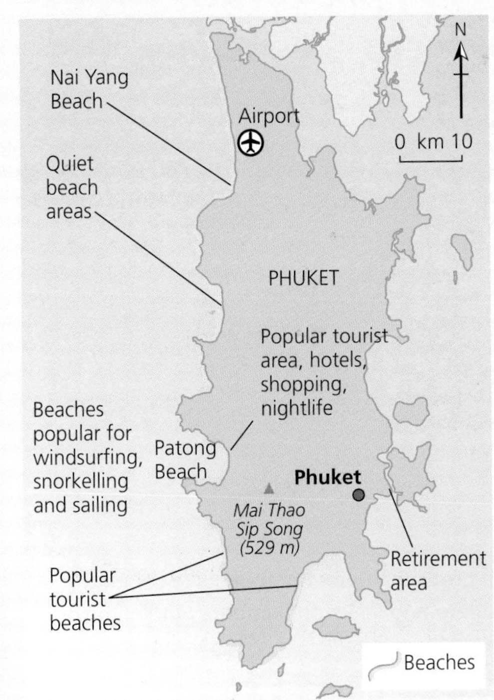

Tourism in extreme environments

Extreme environments are those areas that are often difficult to reach and have extremes of landscape or climate. They are mainly mountainous, cold, as in Antarctica or the Arctic; dry as in deserts; or hot and wet as in the rainforest.

People are attracted to them for their sense of adventure and often the risk-taking that goes with such activities as climbing, diving in Arctic waters, and crossing glaciers.

Adventure holidays in such areas are undertaken in small groups, living 'rough', mainly by young energetic people. More extreme holidays include white-water rafting and mountain paragliding.

Older tourists are also becoming more adventurous and companies now take tourists to remote areas such as the Galapagos Islands, the Maldives and the Gobi Desert.

Case study: *tourism in an extreme environment*

Antarctica

Factfile

- The fifth largest continent.
- Contains 91% of the world's ice.
- Winter temperatures –40 to –70°C.
- Summer temperatures –15 to –35°C.
- The International Antarctic Treaty states that Antarctica should be free from war, pollution and mineral exploitation. There are no international laws governing tourism.

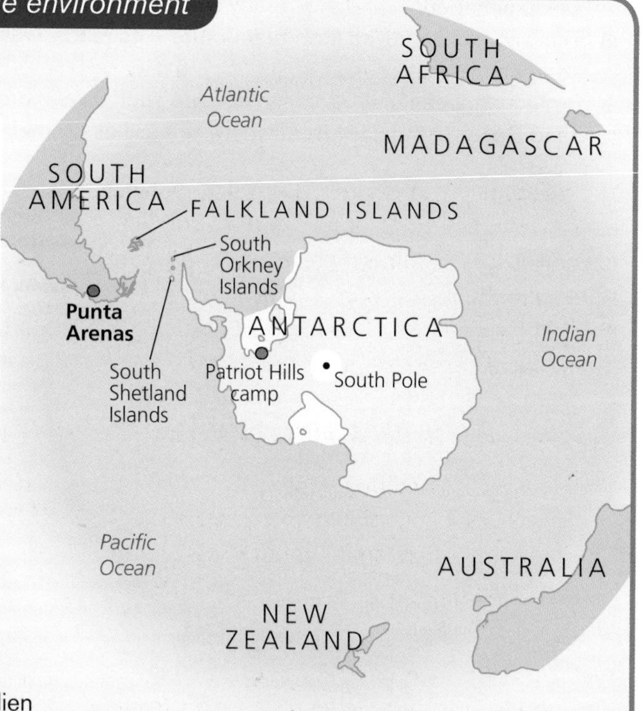

The main attractions are:

- an experience of wilderness
- to view the wildlife and landscape
- activities such as climbing and diving

Tourism is self-regulated:

- dumping waste is prohibited
- guidelines minimise the risk of bringing alien organisms or damaging the environment
- visitors are not allowed to go to sites of special scientific value

The growing number of tourists is causing concern and limits on numbers landing have been set.

Sustainable tourism: ecotourism

Stewardship and conservation

Stewardship is being responsible for looking after the environment for future generations. That means planning to manage the environment sustainably at an international, national and regional level.

For example, at an international level, the Antarctic Treaty seeks to protect the fragile and unique environment with international agreements on wildlife and limits to mineral exploitation. At a national level there is the 'Environmental Stewardship Scheme' in the UK (see page 106).

Conservation entails the protection of the Earth and its resources and involves using its resources in a sustainable way.

Large-scale conservation includes reducing our dependence on fossil fuels and using alternative energy, stopping the destruction of fragile ecosystems such as the rainforest, and protecting special areas such as National Parks.

What is ecotourism?

Ecotourism involves protecting the environment and the way of life of the local people for future generations. Ecotourism aims to:

- reduce to a minimum the impact on the environment
- use small-scale accommodation, which is energy efficient and produces little waste
- use food produced locally to support the local economy
- use local people as guides to provide jobs

Ecotourism and sustainable development

The diagram shows that ecotourism can contribute to sustainable development as tourist numbers are small and it is sensitive to the local culture of the people and the environment.

But this is on a very small scale – the big question is how can the lessons learnt from ecotourism be applied to mass tourism with over 800 million people worldwide taking holidays abroad each year.

Is the answer to encourage ecotourism in particularly sensitive areas and allow mass tourism to continue in less sensitive areas?

Characteristics of ecotourism

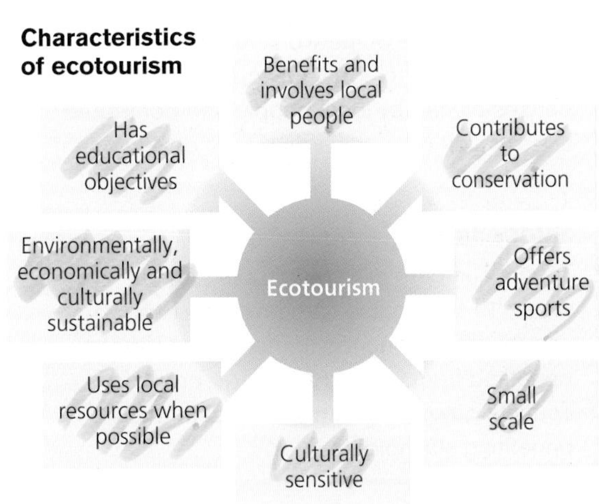

Case study: *ecotourism*

Lushoto, Tanzania

Factfile

- Lushoto lies in the Usambara Mountains of Tanzania.
- The area has a wide diversity of flora and fauna as well as spectacular scenery.
- Tourists come to appreciate the environment and to hike.

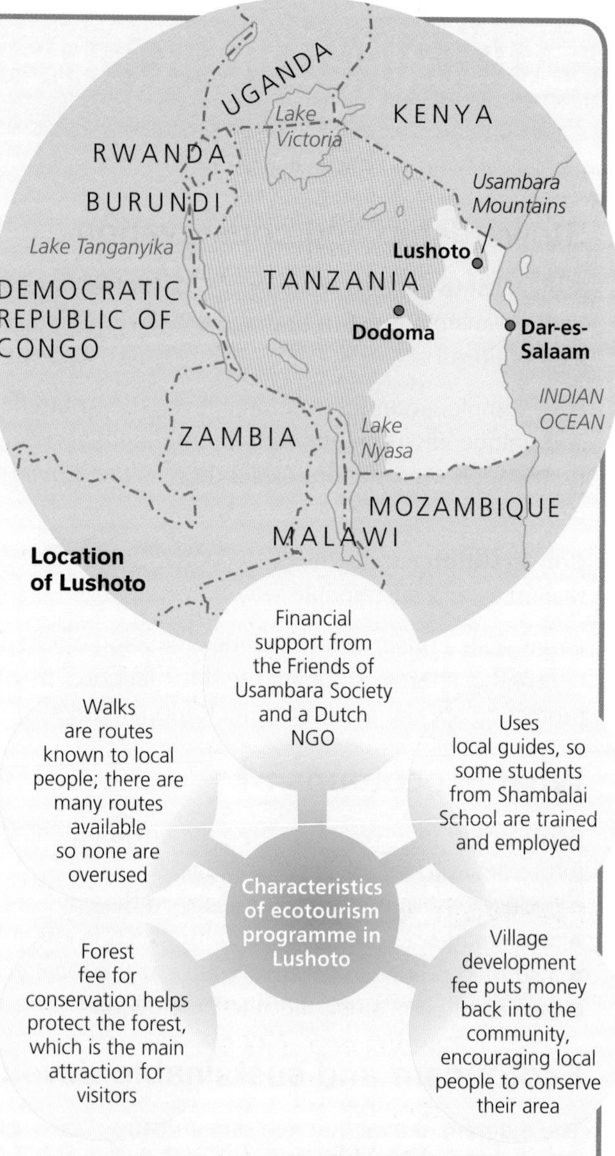
UGANDA
KENYA
Lake Victoria
RWANDA
BURUNDI
Usambara Mountains
Lake Tanganyika
Lushoto
TANZANIA
DEMOCRATIC REPUBLIC OF CONGO
Dodoma
Dar-es-Salaam
INDIAN OCEAN
ZAMBIA
Lake Nyasa
MOZAMBIQUE
MALAWI
Characteristics of ecotourism programme in Lushoto
Financial support from the Friends of Usambara Society and a Dutch NGO
Uses local guides, so some students from Shambalai School are trained and employed
Village development fee puts money back into the community, encouraging local people to conserve their area
Forest fee for conservation helps protect the forest, which is the main attraction for visitors
Walks are routes known to local people; there are many routes available so none are overused

Location of Lushoto

Ecotourism project

The project is run by local people with the support of a Dutch non-governmental organisation and the Tanzanian Tourist Board. It aims to be:

- sensitive to the local environment
- sensitive to the local culture
- benefit local communities by helping to improve living standards and services such as health care and education

The organisers hope that when the local people see the benefits of ecotourism they will be encouraged to protect their environment.

The tours

Tourists are able to experience the local culture, see how people live and work, and enjoy the local environment. Tours last from 34 hours to 3–4 days, mainly between October and August. Because of the seasonal nature of the work, local guides usually have other jobs. Visitors stay in guest houses owned by local people and contribute a fee, which goes towards village development projects.

Successes

Tourist numbers have quadrupled in four years: 14 local guides are employed and communities along the tourist routes have benefited from funding for village projects. Tourists buy goods from the local market benefiting stallholders.

But:

Tourist numbers fell following terrorist attacks in the USA and Kenya and the area is facing competition from other ventures from private companies offering similar experiences. Tourist numbers vary from month to month from 70 or 80 in December and January to over 300 per month in July and August.

Key words

stewardship
conservation
ecotourism

Test yourself

1 **Underline the *five* words or phrases printed in bold in the following extract which are the main reasons for the rise in international tourism in the past 50 years.**

There **are more historic buildings** than there used to be for people to visit. **Travel has become quicker and easier** during the past 50 years. **Cities are larger** so there is more room for hotels for tourists. People in richer countries have **more money to spend** and have **more days of paid leave** so can afford to take more holidays. **People live longer** so can earn more money to spend on travel. With **global warming** there are more sunny days for people to spend on the beach. People **know more about places** in the world which encourages them to travel. Organised trips like **package holidays** make travel easier.

2 **Match the following terms with the correct definition:**

Mass tourism	areas which are difficult to reach with extremes of climate and environment
Honeypot site	being responsible for looking after the environment for future generations
Extreme environment	the protection of the earth and its resources
Ecotourism	large numbers of tourists travel to one destination creating a concentration of hotels and facilities
Stewardship	sustainable tourism which respects the environment
Conservation	a place to which many tourists are attracted causing congestion and overcrowding.

Exam tip

Read each question carefully and be clear what the examiner is asking you to write about. In this question, for example, you need to be able to distinguish clearly between 'extreme environments' and 'ecotourism'.

Examination question

Foundation tier:

(a) For one extreme environment you have studied:

(i) Name the area and explain why tourists visit this area. *(2 marks)*

(ii) What strategies is the area using to cope with large numbers of tourists. *(3 marks)*

(b) For one example of ecotourism you have studied:

(i) Name the area and describe how the example you have chosen is trying to support the local economy. *(3 marks)*

(ii) Describe how the example you have chosen is trying to protect the environment. *(3 marks)*

Higher tier:

(c) Explain what strategies one extreme environment you have studied, which attracts tourists, has put in place to cope with increased numbers of tourists. *(5 marks)*

(d) For one example of ecotourism you have studied explain how this is developing a sustainable form of tourism, which benefits local people and protects the environment. *(6 marks)*

Key word index